STUDENT HANDBOOK

A GUIDE TO CONCEPTS AND PROBLEM SOLVING

PREPARED BY

HARRY NICKLA
Creighton University, Omaha

THIRD EDITION

CONCEPTS OF GENETICS

William S. Klug
Trenton State College at Hillwood Lakes

Michael R. Cummings
University of Illinois, Chicago

Macmillan Publishing Company
New York

Collier Macmillan Canada, Inc.
Toronto

Maxwell Macmillan International Publishing Group
New York   Oxford   Singapore   Sydney

Macmillan Publishing Company
866 Third Avenue, New York, New York 10022

Collier Macmillan Canada, Inc.

Printing: 1 2 3 4 5 6 7   Year: 1 2 3 4 5 6 7

ISBN 0-02-364795-7

Contents

Introduction

Purpose of this book

The intent of this book is to help you understand introductory genetics as presented in **Concepts of Genetics** (3rd edition) by Klug and Cummings. To do so, you must build the conceptual framework in which to place various experiments, examples, and illustrations which are prevalent in genetics and which usually represent the basis of specific homework problems and test questions. To succeed, you must be able to recognize from where in that framework each particular test question is drawn, and what examples are likely to pertain to each concept.

A first course in genetics can be a humbling experience for many students. It is possible that the lowest grades received in ones major, or even in ones undergraduate career, may be in genetics. It is not unusual for some students to become frustrated with their own inability to succeed in genetics. This frustration is felt by most teachers as they field most of the following student comments:

Student frustrations

"I studied all the material but failed your test."

"I must have a mental block to it. I just don't get it. I just don't understand what your are asking."

"I know all the material, but I can't take your tests."

"Where did you get that question? I didn't see anything like that in the book or in my notes."

"This is the first test I have **ever** failed."

"I helped three of my friends last night and I got the lowest grade."

"I am getting a "D" in your course and I have never received less that a "B" in my whole life."

"I stayed up all night studying for your exam and I still failed."

Similar to Algebra

Think back to the first time you encountered "word problems" in your first algebra class. How many times did you say to yourself, your parents, or to your teacher,

> "I hate word problems, I just can't understand them, and why do I need to learn this anyway, I'll never use it."

At that time you had two choices, drop out and be afraid of problem-solving for the rest of your life (which unfortunately happens too often) or regroup, seek help, strip away distractions, and focus in on learning something new and powerful. Because you are taking genetics, you probably succeeded in algebra, perhaps with difficulty at first, and you will probably succeed in genetics.

You were forced in algebra to convert something real and dynamic (two trains leaving at different times from different stations at different speeds - when do they meet?) to a somewhat abstract formula which can be applied to an infinite

Train A → New York   ← Train B Chicago

number of similar problems. In genetics you will again learn something new. It will involve the conversion of something real and dynamic (genes, chromosomes, hereditary elements, gamete formation, gene splicing, and evolution) to an array of general concepts (similar to mathematical formulas) which will allow you to predict the outcome of an infinite number of presently known and yet to be discovered phenomena relating to the origin and maintenance of life.

Mental pictures and symbols

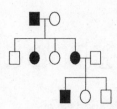

When working almost any algebra word problem it is often helpful to make a simple drawing which relates, in space, the primary participants. From that drawing one can often predict or estimate a likely outcome. A mathematical formula and its solution provides the precise outcome. In understanding genetics it is often helpful to make drawings of the participants whether they be crosses (*Aa* X *Aa*), gametes (*A* or *a*), or the interactions of molecules (anticodon with codon).

As with algebra, the symbolism used to represent a multitude of structures, movements, and interactions, is abstract, informative, and fundamental to understanding the discipline. It is the set of symbols and the relationships among symbols which comprise the universal set of concepts (paradigms) which, as specific examples, make up the framework of genetics. Test questions and problems which, as examples, may be completely unfamiliar to the student, nevertheless refer directly to the basic concepts of genetics.

Concepts of Genetics

Part One: Heredity and Phenotype

Part Two: Molecular Basis of Heredity

Part Three: Organization and Regulation of Genetic Information

Part Four: Genetics of Organisms and Populations

The genetics instructor is competent

Teachers usually adjust the level of a course, the selection of a text, test questions, and lecture material, on two criteria:

> (1) the capability of the students which is determined by factors such as the course prerequisite pattern, and the entry standards of the institution, and

> (2) the experience and expertise of the instructor.

While the instructor may do his/her best to present the material clearly, the burden is on the student to learn genetics. Regardless of how hard a teacher tries to explain certain concepts, the student must be an active participant in the learning process. A student can not enter a genetics classroom, forgetting all that was learned in the prerequisite course(s), and expect the instructor to start from scratch and teach them genetics.

There are interesting, important, and somewhat complex concepts to learn. Such learning requires a focused effort not only by the instructor but also by the student. The student should expect that a competent instructor is attempting to teach them how life, with its variation and constancy, is passed from one generation to the next. The instructor expects that students are responding with a thoughtful, mature, disciplined effort to learn. Anything short of those expectations is likely to result in disappointment on the part of the instructor and the student.

Granted, there may be idiosyncrasies of a given instructor which may be distracting or even annoying to students. But mature students are able to dismiss surface distractions and focus on the subject material. They are unwilling to let an individual stand between them and their right to an understanding of significant biological information.

The tests are fair

Because many students may fail a particular test, it is no indication that the test is not fair. It is expected that some students will have a more difficult time with certain concept areas than others and as a result will do poorly on certain tests. It would be a tremendous disservice to our students if instructors "watered down" or omitted difficult material so that more students would receive better grades. For a variety of reasons, there are temptations (student evaluations, fewer enemies) to give high grades and fail few.

Experienced teachers recognize conceptual areas which are the most difficult for students and often attempt different teaching strategies or hold extra review sessions for students when those areas are encountered. It is up to the student to take advantage of the instructor's offer to help. It is important for the student to be confident that the instructor is competent and that the tests are designed to properly evaluate what the student understands.

There is considerable uniformity in the instruction of genetics. Instructors expect students to learn certain fundamental concepts before passing grades are awarded. Students should expect little compromise in this aspect of evaluation.

It is likely that a student will encounter a test question which involves an unfamiliar example or situation. Genetics teachers may use literally thousands of ways to test a student's understanding of a simple dihybrid cross. Students who do well in genetics are able to focus on understanding concepts, rather than memorizing a multitude of examples.

Attendance and attention are mandatory

Many professors do not take attendance in lectures, therefore, it is likely that some students will opt to take a day off now and then. Unless those students are excellent readers and excellent students in general, continual absences will ususally result in failure.

Remember how difficult it was to set-up and understand the first algebra word problem on your own. It is likely that your ultimate source of understanding came from the course instructor. While using the text is important in your understanding of genetics, the teacher can walk you through the concepts and strategies much more efficiently than a text because a text is organized in a *sequential* manner. A good teacher can "cut and paste" an idea from here and there as needed. To benefit from the wisdom of the instructor, the student must concentrate during the lecture session rather than sit, passively taking notes, assuming that the ideas can be figured out at a later date. Too often the student will not be able to relate to notes passively taken weeks before.

It is necessary for students to attend class and take advantage of the instructor's insights during the lecture sessions. The instructor will not be able to cover all the material in the text. Parts will be emphasized while other areas may be omitted entirely. Since it is the instructor who writes and grades the tests, who is in a better position to prepare the students for those tests?

There is no magic formula for understanding genetics or any other discipline of significance. Learning anything, especially at the college level, requires time, patience, and confidence. First, a student must be willing to focus on the subject matter for an hour or so each day over the entire semester (quarter, trimester, etc.). Study time must be free of distractions and pressured by the presence of clear, realistic goals.

The student must be patient and disciplined. It will be necessary to study when there are no assignments due and no tests looming.

The majority of successful students are willing to read the text ahead of the lecture material, spend time thinking about the concepts and examples, and work as many sample problems as possible. They study for a period of time, stop, then return to review the most difficult areas. They do not try to cram information into marathon study sessions a few nights before the examinations. While they may get away with that practice on occasion, more often than not, understanding the concepts in genetics requires more mature study habits and preparation.

Perhaps a different way of thinking

Because the acquisition of problem-solving ability requires that students rely on new and important ways of seeing things rather than memorizing the book and notes, some students find the transition more difficult than others. Some students are more able to deal in the abstract, concept-oriented framework than others.

Students who have typically relied on "pure memory" for their success, will find a need to focus on concepts and problem-solving. They may struggle at first just as they may have struggled with the first word problem in algebra. But the reward for such struggle is intellectual growth.

Students should expect to grow intellectually in a genetics course and with such growth will come an increased ability to solve a variety of problems beyond genetics. Problem-solving is a process, a style, which can be applied to many disciplines. Few people are actually born with the touch of synthetic brilliance. It comes from probing deeply in a few areas to see how one "gets to the bottom" of things. Then, because the "path is universal" it is easier to know when the bottom has or has not been reached in all disciplines.

At each particular point in a lifetime one has different capabilities. Certain intellectual areas are less developed than others, perhaps because of a particularly simulating teacher in grade school or a particularly poor teacher. It may be a neighbor or a relative that gave you, by chance or by insight, the stimulation (in the form of a particular book or an extended conversation or explanation) that helped you take a mental leap. There is often the misguided impression that if you work hard enough, you can accomplish anything. That is simply not true in all ones endeavors. By working hard in a focused, concentrated fashion, one can accomplish a great many things perhaps the most important of which is the confidence that comes from achieving goals, even if small at first. Hard work, in combination with intellectual maturity, often leads to academic success.

How to study

Genetics is a science which involves symbols (*A, b, p*), structures (chromosomes, ribosomes, plasmids), and processes (meiosis, replication, translation) which interact in a variety of ways. Models describe the manner in which hereditary units are made, how they function, and how they are transmitted from parent to offspring. Because many parts of the models interact both in time and space, genetics can not be viewed as a discipline filled with facts which should be memorized. Rather, one must be, or become, comfortable with

Time,

Work,

Patience

seeking to understand not only the components of the models but also how the models work.

One can memorize the names and shapes of all the parts of an automobile engine, but without studying the interrelationships among the parts in time and space, one will have little understanding of the real nature of the engine. It takes time, work, and patience to see how an engine works and it will take time, work, and patience to understand genetics. And just as one is likely to be fascinated with the movement and power of an engine, such fascination is quite likely to come with an understanding of genetics.

Don't cram. A successful tennis player doesn't learn to play tennis overnight; therefore, you can't expect to learn genetics under the pressure of night-long cramming. It will be necessary for you to develop and follow a realistic study schedule for genetics as well as the other courses you are taking. It is important that you focus your study periods into intensive, but relatively

Study when there are no tests

short sessions each day throughout the entire semester (quarter, trimester). Because genetics tests often require you to think "on the spot" it is very important that you get a good night's sleep before each test. Avoid caffeine in the evening before the test because a clear, rested, well-prepared mind will be required.

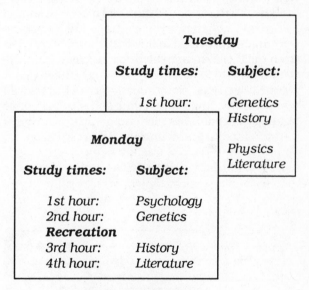

Tuesday	
Study times:	**Subject:**
1st hour:	Genetics History
	Physics Literature

Monday	
Study times:	**Subject:**
1st hour:	Psychology
2nd hour:	Genetics
Recreation	
3rd hour:	History
4th hour:	Literature

Develop a realistic schedule

Study goals. The instruction of genetics is often divided into large conceptual units. A test usually follows each unit. It will be necessary for you to study genetics on a routine basis long before each test. To do so, set specific study goals. Adhere to these goals and don't let examinations in one course interfere with the study goals of another course. Notice that each course being taken is handled in the same way — study ahead of time and don't cram.

Develop a monthly plan

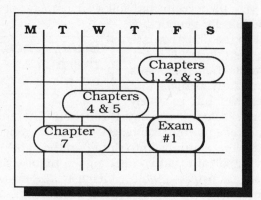

Read ahead. You have been told that it is important to read the assigned material before attending lectures. This allows you to make full use of the information provided in the lecture and to concentrate on those areas which are unclear from the readings. An opportunity is often provided for asking questions. Your questions will be received much more favorably if you can say that after reading the book and listening to the lecture a particular point is still unclear. It is very likely that your question will be quickly dealt with to your benefit and the benefit of others in the class.

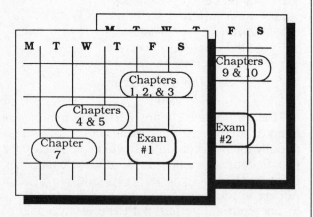

Develop a plan for
the semester

Work the assigned problems. The basic concepts of genetics are really quite straightforward but there are many examples which apply to these models. To help students adjust to the variety of examples and approaches to the models, instructors often assign practice problems from the back of each chapter. If your instructor has assigned certain problems, finish working them *at least* one week before each examination. Before starting with a set of problems, read the chapter carefully and consider the information presented in class.

Suggestions for working problems:

---

(1) work the problem *without* looking at the answer,

(2) check your answer in this book,

(3) if incorrect, work the problem again,

(4) if still incorrect, you don't understand the concept,

(5) re-read your lecture notes and the text,

(6) work the problem again,

(7) if you still don't understand the solution, mark it, and go to the next problem.

In your next study session, return to those problems which you have marked. Expect to make mistakes and learn from those mistakes. Sometimes what is difficult to see one day may be obvious the next day. If you are still having problems with a section, schedule a meeting with your instructor. Usually the problem can be cleared up in a few minutes.

You will notice that in this book, I have presented the solution to each problem. I provide different ways of looking as some of the problems. Instructors often take a problem directly from those at the end of the chapters or they will modify an existing probem. Reversith the "direction" of a question is a common approach. Instead of giving characeristics of the parents and asking for characteristics of the offspring, the question may provide characteristics of the offspring and ask for particulars on the parents.

Separate examples from concepts. As mentioned earlier, genetics boils down to a few (perhaps 15 to 20) basic concepts; however there are many examples which apply to those concepts. Too often students have trouble separating examples from the concepts. Notice that in the "Sample Test" section in this book, I have made such separations clear. Examples allow you to picture, in concrete terms, various phenomena but they don't exemplify each phenomenon or concept in its entirety.

Caution in the use of old examinations. Often it is customary for students to request or otherwise obtain old examinations from previous students. Such a practice is loaded with pitfalls. First, students often, albeit unconsciously, find themselves "second guessing" about questions on an upcoming examination. They forget that an examination usually only tests over a subset of the available information in a section. Therefore entire "conceptual areas" may be available which have not appeared on recent exams.

Often the reproductions of old examinations are of poor quality (having been copied and passed around repeatedly) and it is difficult to determine whether the answer provided is the correct one or if it was incorrect and marked wrong. In addition, if a question has the same general structure as one on a previous examination, but is modified, students often provide an answer for the "old" question rather than the one being asked.

Granted, it is of value to see the format of each question and the general emphasis of previous examinations, but remember that each examination is potentially a new production capable of covering areas which have not been tested before. This is especially likely in a course such as genetics where the material changes very rapidly. Don't try to figure out what will be asked. Study all the material as well as possible.

Structure of this book

The intent of this book is to help you understand the concepts of genetics as given in the text and most likely in the lectures, then to apply these concepts to the solution of all problems and questions at the ends of the chapters. Rather than merely provide you with the solutions to the problems I have tried to walk you through each component of each question so that you can see where information is obtained and how it can be applied in the solution.

Vocabulary: Organization and listing of terms. Understanding the vocabulary of a discipline is essential to understanding the discipline. Throughout the text by Klug and Cummings you will find terms in bold print. Such terms generally refer to structures or substances, processes/methods, and concepts. I have separated these terms and *other important terms* into these categories.

> ***Structures and Substances***
>
> ***Processes/Methods***
>
> ***Concepts***

Those terms or concepts which require special explanation or are more complex or intimately related to other terms are denoted with a code (F2.1, F23.2, etc.) which refers you to the figures immediately following the ***Concepts*** section.

You may see a term present in several categories or in categories other than those you would envision. Since categorization *per se* is of little significance, don't worry about category location. Use the listings as checklists to make certain that you understand the meaning of each term in each chapter. Also, by a given term's category, you can begin to understand whether it refers to a structure or substance, a process or method, or a more general concept. Notice that the various terms are not redefined. It is important that you use the Klug and Cummings text for the original definitions.

Concepts. In the section *"Vocabulary: Organization and Listing of Terms"* you will find a section called *"Concepts"* after which there may be a simple sketch or two to help you focus a particular concept. Such sketches are oversimplifications and you should fill in the details by examining the textbook and the lecture notes.

Solved problems. Each of the problems at the end of each chapter is solved from a beginner's point of view. There are other features of this section. Most of the answers to the questions and problems will refer you to specific sections, usually specific tables and figures, of the Klug and Cummings (K/C) text. Be certain that you fully understand the solution to each of the questions suggested or assigned by your instructor.

Supplemental questions. A series of solved sample test questions supplement the questions provided in the text and help you determine your level of preparation. These sample test questions are located at the end of each of the major units. Concepts relating to each question as well as common errors are presented in boxes before and after each answer.

> **Understand the words and phrases of the discipline**

> **Supplemental Questions**
>
> **Concepts**
>
> **Comprehensive Solution**
>
> **Common errors**

1

An Introduction to Genetics

Vocabulary: Organization and Listing of Terms

Historical

Prehistoric times

    domesticated animals

    cultivated plants

Greek Influence

    Hippocrates

    Aristotle

    Lamarck (1809) - **Philosophie Zoologique**

      pangenesis

      acquired characteristics

    procreation

      religious beliefs

      special creation

      Harvey

        epigenesis

    preformation

      homunculus

        ovists

        spermists

      spontaneous generation

Fixity of species

    Linnaeus

    Kolreuter

      hybridization

      parental types

      backcross

    Naturphilosophie

Evolution

    Darwin (1859)

      The Origin of Species

      natural selection

      pangenesis

Genetics

    Mendel (1865)

    Rediscovery (1900)

        Correns, de Vries, Von Tschermak

        Bateson

Soviet Genetics

    Lysenko

        vernalization

    Vavilov

Structures and Substances

Center for heredity

    nucleus

    nucloid region

    viral head

Genetic material

    DNA (deoxyribonucleic acid)

    RNA (ribonucleic acid)

        nitrogenous bases

    genes

    chromosomes

        chromatin fibers

Associated substances/structures

    amino acids

messenger RNA

ribosome

ribosomal RNA

transfer RNA

enzymes

Processes/Methods

Mitosis

Meiosis

Transmission genetics

    pedigree analysis

    cytological investigations

    chromosome theory of inheritance

    karyotypes

Molecular and biochemical analysis

    recombinant DNA technology

Genetic structure of populations

Basic research

Applied research

    agriculture

        "Green Revolution"

        Borlaug

    medicine

        genetic counseling

        human genetic engineering

Concepts

Genetic information

    stored

    altered

    expressed

    regulated

Genetics

    heredity

    variation

        alleles

Diploid number (2n) (F1.1)

    polyploid

    haploid number (F1.1)

    homologous chromosomes (F1.2)

Genetic variation

    gene mutations

    chromosomal aberrations

Genetic information

    genetic code

    triplet

    transcription

    translated (translation)

F1.1. Below is pictured the union of two gametes, each with the *n* chromosome number. The haploid chromosome number (n) is usually found in gametes while the diploid chromosome number (2n) is usually found in the zygote and somatic tissues of the individual. There are higher levels of "ploidy" such as 3n, 4n, and so on.

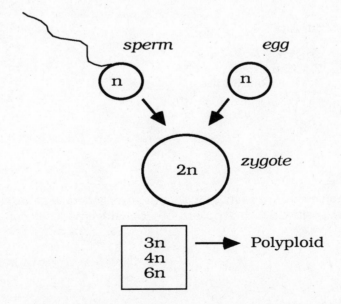

F1.2. One of the most important concepts to be mastered involves an understanding of homologous chromosomes. Each parent normally contributes one of each type of chromosome to the zygote. Since there are two gametes which unite to form the zygote, there must be two of each type of chromosome in each zygote. These two chromosomes of the same type are called homologous chromosomes or homologous pairs of chromosomes. In humans we have 23 pairs of homologous chromosomes. One of each pair came from our mother, the other member of each pair came from our father (as shown in F1.1).

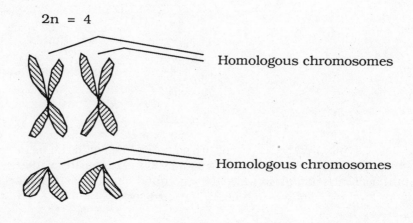

2n = 4

Homologous chromosomes

Homologous chromosomes

Criteria:

1. Size
2. Centromere position
3. Other similarities

Homologous chromosomes are similar in that they are generally of the same size, have the same centromere position, and have many other characteristics in common.

Solutions to Problems and Discussion Questions

1. Both were concerned with subjects of the reproduction, heredity, and origin of humans, and the shifting of interest from religious mythology to philosophical and scientific inquiries. Hippocrates argued that male semen is formed in various parts of the body (healthy or diseased) and transported through blood vessels to the testicles. Such "humors" carried the hereditary traits. Thus the theory of pangenesis was formed. Aristotle was critical of pangenesis because it did not explain the appearance of features which skipped generations. Aristotle suggested that semen contained a vital heat which could produce offspring in the form of the parents.

2. *Pangenesis* refers to a theory that various parts of the body contain "humors" which bear the hereditary traits and gather in the reproductive organs. *Epigenesis* refers to the theory that organisms are derived from the assembly and reorganization of substances in the egg which eventually lead to the development of the adult. *Preformationism* is a 17th century theory which states that the sex cells (eggs or sperm) contain miniature adults, called homunculi, which grow in size to become the adult.

3. Darwin was aware of the physical and physiological diversity of members within and among various species. He was aware that varieties of organisms could be developed through selective breeding (domestication), that a species is not a fixed entity, and that while certain groups of organisms could be hybridized, other groups could not. He was aware of conflicts between religious views and the fossil record. He understood geology, geography, and biology and that organisms tend to leave more offspring than the environment can support.

4. Darwin's theory of natural selection proposed that more offspring are produced than can survive, and that in the competition for survival, those with favorable variations survive. Over many generations, this will produce a change in the genetic make-up of populations if the favorable variations are inherited. Darwin did not understand the nature of heredity and variation which led him to lean toward older theories of pangenesis and inheritance of acquired characteristics.

5. DNA is composed of polymers of nucleotides which can exist either in either single-stranded or double-stranded forms. In non-viral systems, DNA is usually double-stranded. DNA and in some cases proteins associate to form chromosomes. During the cell cycle, DNA is duplicated by a variety of enzymes so that daughter cells inherit copies of the parental DNA. Sections of DNA which produce some influence on the organism are called genes. Genes usually exert their influence by producing proteins through the process of transcription and translation. These topics will be described in considerable detail in Part Two of the K/C text. A simplified diagram of these phenomena is given below:

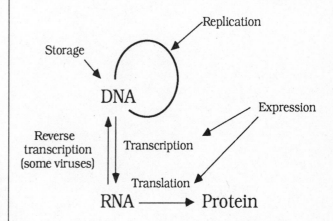

6.   *Transmission* genetics is the most classical approach in which the patterns of inheritance are studied through selective matings or the results of natural matings.   Mendel observed results from precisely defined matings and provided models based on transmission genetics.

A second approach involves physical, traditionally microscopic (light and electron), examination of chromosomes.   With the discovery of mitotic and meiotic processes, and the knowledge that genes are located on chromosomes,   much interest centers on the *cytological investigation* of chromosomes.

Molecular and biochemical analysis of the genetic material has recently evolved into one of the most exciting and rapidly growing subdisciplines of genetics.   Originating in the early 1940s with studies of bacteria and viruses, much information as to the nature of gene expression, regulation, and replication has been provided.   *Recombinant DNA technology* has had a significant impact in this area as well as others.

In *population* genetics the interest is in the behavior of genes in groups of organisms (populations) often with an interest in the factors which change gene frequencies in time and space. Hence population geneticists are often interested in the process of evolution.

7. *Basic* research involves the study of the fundamental mechanisms of genetics as described in the answer to question #6 above.   *Applied* research makes use of the information provided by basic research. Many applications in agriculture and medicine are described in the text.

8.   For over twenty years, Lysenko directed Soviet genetics based on the idea of inheritance of acquired characteristics; that plant productivity could be improved in an inherited fashion, by changes in the environment.   Such was in line with Marxist political thought.   Geneticists in the United States were firmly entrenched in Mendelian interpretations and experiencing a "golden age" of genetics with the birth of molecular biology.

9.   Norman Borlaug applied Mendelian principles of hybridization and trait selection to the development of superior varieties of wheat. Such varieties are now grown in many countries, including Mexico, and have helped maintain the world supply of food.   This change in worldwide agricultural food production has been called the "Green Revolution."

10.   In the last 40 years human transmission, cytological, and molecular genetics have provided an understanding of many human diseases. There is promise that a certain amount of human suffering will be minimized by the application of genetics to medicine.   Major areas of activity include genetic counseling, gene mapping and identification, disease diagnosis, and genetic engineering.

2

Mitosis and Meiosis

Vocabulary: Organization and Listing of Terms

Structures and Substances

Cells

    plasma membrane

    cell wall

        peptidoglycan

    capsule

    cell coat

        ABO antigens

        MN antigens

        histocompatibility antigens

    nucleus, nucleoid

        genetic material

        chromatin

        chromosomes

            genes

            nucleolus

            nucleolar organizer

    cytoplasm

        endoplasmic reticulum

        ribosomes

        cytoskeleton

        mitochondria

        centrosome

        basal body

        centrioles

        spindle fibers

        cytosol

Chromosomes

    centromere

    metacentric

    submetacentric

acrocentric

telocentric

    karyotype

        sex-determining chromosomes

Mitosis

    zygotes

        centrosome

        spindle fibers

        chromatid

        sister chromatid

            cell plate

            middle lamella

            cell furrow

Meiosis

    bivalent

    tetrad

    dyad

    monad

        chiasma (chiasmata)

        synaptonemal complex

        chromomeres

Gametogenesis

    spermatogonium

    primary spermatocyte

    secondary spermatocyte

spermatid

spermatozoa (sperm)

oogonium

primary oocyte

secondary oocyte

first polar body

ootid

second polar body

ova (ovum)

Processes/Methods

Oxidative phases of cell respiration

Photosynthesis

Mitosis

    cell cycle

    cytokinesis

    interphase

    S phase

    G_1 and G_2

        prophase

        prometaphase

        metaphase

        anaphase

        telophase

Regulation

    significance

    cytoplasmic influences

        critical signals

        initiation of DNA synthesis

        G_0

        R point

        G_1, G_1 to S transition

        oncogenes

        protooncogenes

Meiosis

    reductional

    equational

    prophase I

        leptonema

        zygonema(synapsis)

        pachynema

        diplonema

        diakinesis (terminalization)

    metaphase I

    anaphase I

        disjunction

    telophaseI

    prophase II

    metaphase II

    anaphase II

    telophase II

Spermatogenesis

Spermiogenesis

Oogenesis

    first meiotic division

    second meiotic division

Nondisjunction

    monosomy

    trisomy

Sexual reproduction

    reshuffle chromosomes

    provides for crossing over

Methods

    light microscopy

    transmission electron microscopy

    autoradiography

Concepts

Homologous chromosomes (F2.2)

    diploid number(F2.1)

    loci(F2.2)

    haploid genome (haploid number, n) (F2.2)

    alleles (F2.2)

Mitosis (T2.1)

   identical daughters

   equivalent genetic information

Meiosis (F2.3)

   produces gametes or spores

   reshuffles genetic combinations
     (chromosomes)

   genetic recombination (crossing over))

   constant amount of genetic material

   production of variation

Fertilization

   reconstitution of genetic material

Segregation

Independent assortment

Primary nondisjunction

Secondary nondisjunction

Gametophyte stage

Sporophyte stage

Sexual reproduction

F2.1. Diagram showing the relationships among stages of interphase chromosomes, chromosome number, and chromosome structure in an organism with a diploid chromosome number of 4 (2n = 4). There are two pairs of chromosomes, one large metacentric, one smaller metacentric. Individual chromosomes can not be seen at interphase, therefore, the drawings represent chromosomes *as if* they could be individually viewed.

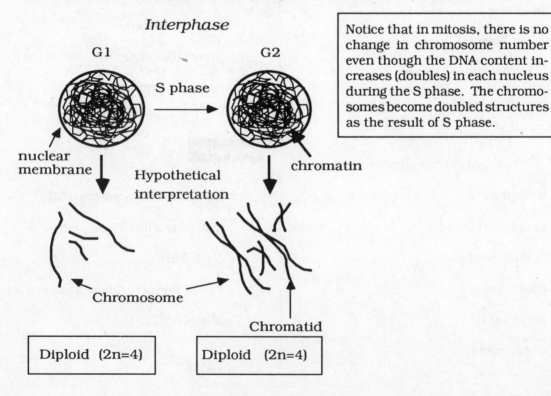

Notice that in mitosis, there is no change in chromosome number even though the DNA content increases (doubles) in each nucleus during the S phase. The chromosomes become doubled structures as the result of S phase.

F2.2. Important nomenclature referring to chromosomes and genes in an organism where the diploid chromosome number is 4 (2n = 4). There are two pairs of chromosomes, one large metacentric, and one smaller metacentric. Sister chromatids are *identical* to each other while homologous chromosomes are *similar* to each other in terms of overall size, centromere position, function, and other factors as described in your text.

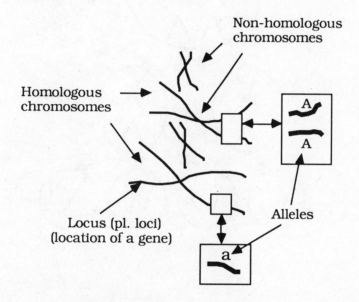

T2.1. Illustration of the relationship between chromosome number and stages of the mitotic cycle.

Cell cycle stage	Chromosome number	
	humans	**fruit flies**
Interphase	2n = 46	2n = 8
G1	2n = 46	2n = 8
S	2n = 46	2n = 8
G2	2n = 46	2n = 8
Mitosis		
Prophase	2n = 46	2n = 8
Metaphase	2n = 46	2n = 8
Anaphase	2n = 46	2n = 8
Telophase	2n = 46	2n = 8

F2.3. Illustrations of chromosomes of meiotic and mitotic cells in an organism with a chromosome number of 4 (2n = 4).

Meiosis

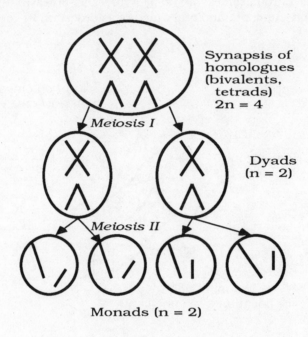

Synapsis of homologues (bivalents, tetrads) 2n = 4

Meiosis I

Dyads (n = 2)

Meiosis II

Monads (n = 2)

Mitosis

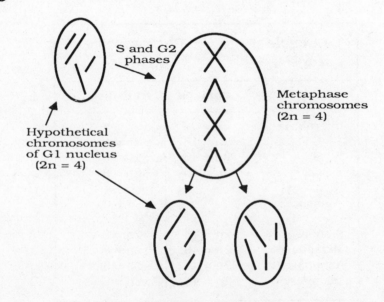

S and G2 phases

Metaphase chromosomes (2n = 4)

Hypothetical chromosomes of G1 nucleus (2n = 4)

Solutions to Problems and Discussion Questions

1.

(a) During interphase of the cell cycle (mitotic and meiotic), chromosomes are not condensed and are in a genetically active, spread out, form. In this condition, chromosomes are not visible as individual structures under the microscope (light or electron). See F2.1 for a sketch of what *chromatin* might look like. Chromatin contains the genetic material which is responsible for maintaining hereditary information (from one cell to daughter cells and from one generation to the next) and production of the phenotype.

(b) The *nucleolus (pl. nucleoli)* is structure which is produced by activity of the nucleolar organizer region in eukaryotes. Composed of ribosomal RNA and protein, it is the site for the production of ribosomes. Some nuclei have more than one *nucleoli*. Nucleoli are not present during mitosis or meiosis because in the condensed state of chromosomes, there is little or no RNA synthesis.

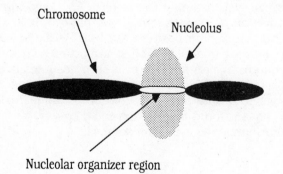

Chromosome

Nucleolus

Nucleolar organizer region

(c) The *ribosome* is the site where various RNAs, enzymes, and other molecular species assemble the primary sequence of a protein. That is, amino acids are placed in order as specified by messenger RNA. Ribosomes are relatively non-specific in that virtually any ribosome can be used in the translation of any mRNA. The structure and function of the ribosome will be described in greater detail in later chapters of the K/C text.

(d) The *mitochondrion (pl. mitochondria)* is a membrane-bound structure located in the cytoplasm of eukaryotic cells.   It is the site of oxidative phosphorylation and production of relatively large amounts of ATP. It is the trapping of energy in ATP which drives many of important metabolic processes in living systems.

(e) The *centriole* is a cytoplasmic structure involved (through the formation of spindle fibers) in the migration of chromosomes during mitosis and meiosis. (K/C-Figs.2.1, 2.7, and 2.9)

(f) The *centromere* serves as an attachment point for sister chromatids (see F2.1) a point where spindle fibers attach to chromosomes. The centromere divides during mitosis and meiosis II thus aiding in the partitioning of chromosomal material to daughter cells. (K/C-Figs.-2.3, 2.7, 2.8, and 2.9) Failure of centromeres or spindle fibers to function properly may result in nondisjunction. (K/C-Fig.2.12)

2.   One of the most important concepts to be gained from this chapter is the relationship which exists among chromosomes in a single cell. Chromosomes which are homologous share many properties including:

overall length; look carefully at F2.1 and F2.2, and K/C-Figs.2.4 and 2.9 to see that each cell prior to anaphase I, contains two chromosomes of approximately the same overall length.

position of the centromere (metacentric, submeta-centric acrocentric, telocentric); Again, look carefully at F2.1 and F2.2, and K/C-Figs.2.3, 2.7, and 2.9. Notice that in each if there is one metacentric chromosome, there will be another metacentric chromosome.

banding patterns; Look carefully at K/C-Fig.2.4 and notice how similar the banding patterns are of the homologous (side by side) chromosomes. Look at the two homologous chromosomes of pair #1 for example. While the overall length of chromosome pairs #16 and #17 appear to be the same note the difference in banding patterns of these non-homologous chromosomes. Compare chromosome pairs #19 and #20 for a more striking example.

Notice also that sister chromatids have identical banding patterns as would be expected since sister chromatids are, with the exception of mutation, identical copies of each other. We would expect that homologous chromosomes would have banding patterns which are very similar (but not identical) because homologous chromosomes are genetically similar but not genetically identical.

type and location of genes; Notice in F2.2 that a *locus* signifies the location of a gene along a chromosome. What that really means is that for each characteristic specified by a gene, like blood type, eye color, skin pigmentation, there are genes located along chromosomes. The *order* of such loci is identical in homologous chromosomes, but the genes themselves, while being in the same order, may not be identical.

Look carefully at the inset (box) in the upper right portion of F2.2 and see that there are alternative forms of genes, *A* and *a*, at the same location along the chromosome. *A* and *a* are located at the same place and specify the same *characteristic* (eye color for example) but there are slightly different manifestations of eye color (*brown* vs. *blue* for example). Just as an individual may inherit gene *A* from the father and gene *a* from the mother, each zygote inherits one homologue of each pair from the father and one homologue of each pair from the mother.

autoradiographic pattern; homologous chromosomes tend to replicate during the same time of S phase.

Diploidy is a term often used in conjunction with the symbol *2n.* It means that both members of a homologous pair of chromosomes are present. Refer to F2.1 in this book. Notice that during mitosis, the normal chromosome complement is 2n or diploid. In humans, the diploid chromosome number is 46 while in *Drosophila* it is 8. K/C-Tab.2.1 lists the *haploid* chromosome number for a variety of species. Notice that in man and flies, the haploid chromosome number is one-half the diploid number. This applies to other organisms as well. However, it is very important to realize that *haploidy* specifically refers to the fact that each haploid cell contains *one chromosome of each homologous pair of chromosomes.*

Compare the nuclear contents of a spermatid and a cell at zygonema in K/C-Fig.2.9. Note that each spermatid contains one member of each of the original chromosome pairs (seen at zygonema). Haploidy of usually symbolized as *n.*

The change from a diploid (2n) to haploid (n) occurs during *reduction division* when tetrads become dyads during meiosis I. In the column representing the number of human chromosomes, the primary spermatocyte (2n=46) becomes two secondary spermatocytes each with n = 23.

3. As you examine the criteria for *homology* in question #2 above, you can see that overall length and centromere position are but two factors required for homology. Most importantly, genetic content in non-homologous chromosomes is expected to be quite different. Other factors including banding pattern and time of replication during S phase would also be expected to vary among non-homologous chromosomes.

4. Because much of Chapter #2 deals with meiosis, it would be best to deal with this question by reading the section entitled "Meiosis and Sexual Reproduction" in K/C and examining K/C-Figs.2.9 and 2.11.

Understanding meiosis and all the related terms is essential for an understanding of genetics. There are several sample test questions which will help you determine your understanding of meiosis.

5. The first sentence tells you that $2n = 16$ and it is a question about mitosis. Since each chromosome in prophase is doubled (having gone through an S phase) and is visible at the end of prophase, there should be 32 chromatids. Because the centromeres divide and what were previously sister chromatids migrate to opposite poles during anaphase, there should be 16 chromosomes moving to each pole. If you refer to F2.1 and K/C-Fig.2.7 you will see an example with $2n = 4$ and that there are four doubled chromosomes in prophase. Notice that there are eight chromatids visible at *late* prophase.

6. Refer to K/C-Fig.2.3 for an explanation. Pay special attention to the right-hand portion of the figure and notice the different anaphase shapes. Your understanding of these structures will be determined by several of the sample test questions at the end of this section.

7. Because of a cell wall around the plasma membrane in plants, a cell plate, which was laid down during anaphase, becomes the middle lamella where primary and secondary layers of the cell wall are deposited.

8. Carefully read the section entitled "Mitosis" in K/C. Refer to F2.1 and K/C-Fig.2.5 for information pertaining to the interphase. Refer to K/C-Fig.2.7 for a diagram of mitosis. Notice that, in contrast to meiosis, there is no pairing of homologous chromosomes in mitosis and the chromosome number does not change (see T2.1).

Autoradiography is a technique which can be used to determine that there is DNA synthesis during the interphase. If DNA precursors are available early in interphase, they can be incorporated into replicating DNA during the S phase. Autoradiographs made after the S phase will show radioactive material in the DNA.

9. Compared with mitosis which maintains a chromosomal constancy, meiosis provides for a reduction in chromosome number, and an opportunity for exchange of genetic material from homologous chromosomes (see K/C-Fig.2.9). In mitosis there is no change in chromosome number (see T2.1) or kind in the two daughter cells whereas in meiosis numerous potentially different haploid (n) cells are produced. During oogenesis, only one of the four meiotic products is functional (see K/C-Fig.2.11); however, four of the four meiotic products of spermatogenesis are potentially functional.

10.

(a) *Synapsis* is the point-by-point pairing of homologous chromosomes during prophase of meiosis I (see K/C-Fig.2.9 at the zygonema stage).

(b) *Bivalents* are those structures formed by the synapsis of homologous chromosomes. In other words, there are two chromosomes (four chromatids) which make up a bivalent. If an organism has a diploid chromosome number of 46, then there will be 23 bivalents in meiosis I.

(c) *Chiasmata* is the pleural form of chiasma and refers to the structure, when viewed microscopically, of crossed sister chromatids. Notice in K/C-Fig.2.9 the exchange of chromatid pieces in diplonema and diakinesis.

(d) *Crossing over* is the exchange of genetic material between chromatids. Also called recombination, it is a method of providing genetic variation through the breaking and rejoining of chromatids. Notice in K/C-Fig.2.9 the mixing of genetic material along the length of the chromatids.

(e) *Chromomeres* are patches of chromatin which look different from neighboring patches along the length of a chromosome. They are depicted in K/C-Fig.2.9 at the leptonema stage.

(f) Examine F2.1 in this book. Notice that *sister chromatids* are "post-S phase" structures of replicated chromosomes. Sister chromatids are genetically identical (except where mutations have occurred) and are attached to the same centromere. Identify the sister chromatids in K/C-Fig.2.7

and 2.9.   Note that sister chromatids separate from each other during anaphase of mitosis and anaphase II of meiosis.

(g) Tetrads are synapsed homologous chromosomes thereby composed of four chromatids. There are as many tetrads as the haploid chromosome number. Identify tetrads in K/C-Fig.2.9.

(h) Actually, each tetrad is made of two dyads which separate from each other during anaphase I of meiosis. Dyads are composed of two chromatids joined by a centromere. Identify dyads in K/C-Fig.2.9.

(i)   At anaphase II of meiosis, the centromeres divide and sister chromatids go to opposite poles. Identify monads in K/C-.Fig.2.9.

(j) The *synaptonemal complex* is a nuclear structure which is often found associated with synapsed meiotic chromosomes. It looks and acts like a zipper (K/C-Fig.2.10).

11. Look carefully at F2.3 in this book and notice that for a cell with 4 chromosomes, there are two tetrads each comprised of a homologous pair or chromosomes.

(a) If there are 16 chromosomes there should by 8 tetrads.

(b) Also note that, after meiosis I and in the second meiotic prophase there are as many dyads as there are *pairs* of chromosomes. There will be 8 dyads.

(c) Because the monads migrate to opposite poles during meiosis II (from the separation of dyads) there should be 8 monads migrating to *each* pole.

12. Examine K/C-Fig.2.11 and the figure below. Notice that major differences include the sex in which each occurs, and that the distribution of cytoplasm is unequal in oogenesis but considered to be equal in the products of spermatogenesis. Chromosomal behavior is the same in spermatogenesis and oogenesis except that the nuclear

activity in oogenesis is "off-center" thereby producing first and second polar bodies by unequal cytoplasmic division. Each spermatogonium and primary spermatocyte produces four spermatids whereas each oogonium and primary oocyte produces one ootid.   Because early development occurs in the absence of outside nutrients, it is likely that the unequal distribution of cytoplasm in oogenesis evolved to provide sufficient information and nutrients to support development until the transcriptional activities of the zygotic nucleus begin to provide products. Polar bodies probably represent non-functional by-products of such evolution.

Spermatogenesis

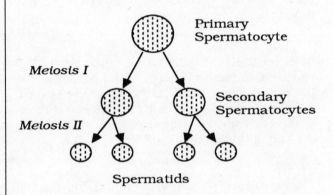

Oogenesis

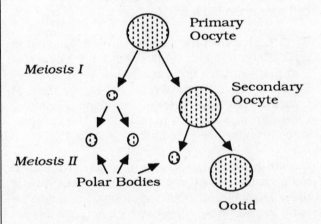

13. This answer contains several parts. First, through independent assortment of chromosomes at anaphase I of meiosis, daughter cells (secondary spermatocytes and secondary oocytes) may contain different sets of maternally- and paternally-derived chromosomes. Examine the diagram below. Notice that there are several ways in which the maternally- and paternally-derived chromosomes may align. Can you calculate the probability of all the maternally-derived chromosomes going to the "right-hand" pole?

Second, crossing over, which happens at a much higher frequency in meiotic cells as compared to mitotic cells, allows maternally- and paternally-derived chromosomes to exchange segments thereby increasing the likelihood that daughter cells (secondary spermatocytes and secondary oocytes) are genetically unique. Notice such exchanges in K/C.Fig.2.9.

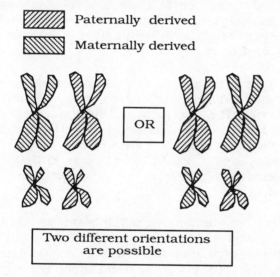

Paternally derived

Maternally derived

OR

Two different orientations are possible

Daughter cells resulting from the process of mitosis are usually genetically identical.

14. Notice the information provided in question #13 above. You can see the different chromosomal alignments (two) possible. In general, there are 2n possible combinations of meiotic products where n is the number of bivalents (number of chromosome pairs which is the same as the haploid number of chromosomes). In this question,

crossing over has provided a situation (very natural indeed) in which the sister chromatids contain pieces of the homologues. This makes each sister chromatid unique so that the random alignment of the chromosomes at metaphase II increases the number of unique gametes possible. Notice in the prophase II and anaphase II portions of K/C-Fig.2.9, a different alignment would produce different products. Given that there are two possible alignments of bivalents at metaphase I and two possible alignments at metaphase II (because crossing over is involved), there are 2 X 2 or four unique combinations of the four spermatids.

15. As you first read this question, think about an animal with n = 6, therefore there will be 6 tetrads. The question concerns one of these tetrads as it passes normally through meiosis I, but one dyad undergoes secondary nondisjunction. As shown in K/C-Fig.2.12, secondary nondisjunction occurs during the second meiotic division. Notice in the second sentence, it says that "the involved dyad ends up intact in the ovum."

(a) The mature ovum should therefore contain n +1 chromosomes. The five chromosomes from normal disjunction and two (from one dyad) from the nondisjunctional chromosome.

(b) The second polar body did not receive one of the six monads it would normally receive, so it should have five monads (which are chromosomes).

(c) When the normal sperm with its n chromosome number combines with an n+1 ovum, it will produce a zygote with 2n +1, or thirteen chromosomes. This condition is termed, *trisomy*.

16. This question specifically tests your understanding of meiosis and the behavior of chromosomes during anaphase. In this question you must first visualize the alignment of the three homologous chromosome pairs C1,C2, M1,M2, and S1,S2 in mitosis (no synapsis of homologues) as compared with the alignment in meiosis.

(a)

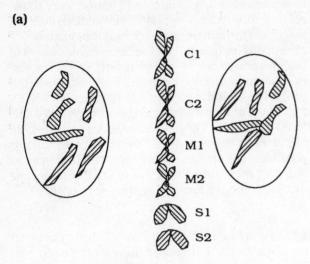

C1

C2

M1

M2

S1

S2

[///] Paternally derived

[\\\] Maternally derived

The two daughter cells will have
the same chromosomes after mitosis

(b)

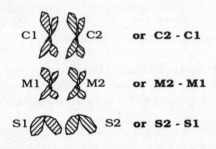

C1   C2   **or C2 - C1**

M1   M2   **or M2 - M1**

S1   S2   **or S2 - S1**

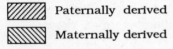

[///] Paternally derived

[\\\] Maternally derived

Notice that there are no constraints
on the alignment of different pairs
of homologous chromosomes, therefore
one could list 8 configurations.

(c) Because of the independent assortment of non-homologous chromosomes at anaphase I and the fact that anaphase II separates sister (identical) chromatids, there will be eight different meiotic (haploid) products.

17.

(a) While it is likely that the molecular processes involved in crossing over occur earlier, *crossing overs* are known to have occurred by pachynema.

(b) *Synapsis* occurs at zygonema (K/C-Fig.2.9) when homologous chromosomes align in a point-by-point fashion to form bivalents or tetrads.

(c) Chromosomes begin to condense at the earliest stage of prophase I, leptonema (K/C-Fig.2.9).

d. *Chiasmata* are clearly visible at diplonema (K/C-Fig.2.9).

18. In angiosperms, meiosis results in the formation of microspores (male) and megaspores (female) which give rise to the haploid male and female gametophyte stage. Micro- and mega-gametophytes produce the pollen and the ovules respectively. Following fertilization, the sporophyte is formed.

19. First, review K/C-Figs.2.5 and 2.6 and notice that interphase is divided into G_1, S, and G_2 phases. The G_0 phase occurs during the G_1 phase and indicates that portion of the cycle when the activity of some cells is arrested. The section entitled 'Regulation of the Cell Cycle' in K/C text identifies the following factors as regulating the cell cycle:

cell type: Meristematic cells in plants and certain epithelial cells in animals pass continuously from one cell cycle to another. Those cells whose reproductive activity is arrested, either permanently or temporarily, usually stop their cycle before the S phase, designated G0. The most variable period is G1.

cytoplasmic influences: Experiments involving fusion of cells which are at different stages of the cell cycle, showed that the cytoplasm contains molecular information which controls DNA synthesis, apparently a critical signal in the cell cycle.

genetic data: While it is likely that the presence of the "division signals" mentioned above is related to the genetic make-up of each cell, experiments by L. Hartwell using yeast, *Saccharomyces cerevisiae,* have shown that many genes are involved in the control of the cell cycle. Called *cdc* mutations, they are responsible for a range of cellular defects including failure to initiate DNA synthesis. Oncogenes (proto-oncogenes) are also known to play specific roles in the regulation of the cell cycle.

20.   It is probable that the factors which cause cancer are related to the same, natural factors which condition the entry of cells into the S phase. Studies mentioned in the previous question provide insight into when regulation is most apparent and what types of molecules are likely to be involved. If these substances can be manipulated then it is likely that the factors which influence cancer may one day be manipulated.

3

Mendelian Genetics

Vocabulary: Organization and Listing of Terms

Historical

Mendelian Genetics (Gregor Mendel)

Pisum sativum

units of inheritance (particulate)

Rebirth (1900)

Hugo DeVries

Karl Correns

Erich Tschermak

Structures and Substances

Unit factors

Genes

Alleles

multiple alleles

Locus

Processes/Methods

Transmission genetics

true-breeding, "bred true"

monohybrid cross

maternal parent

paternal parent

selfing, self-fertilizing

reciprocal cross

test cross

parental generation (P_1)

first filial generation (F_1)

second filial generation (F_2)

ratios

3:1

9:3:3:1

27:9:9:9:3:3:3:1

product law

2^n (n = haploid chromosome number)

sum law

dihybrid cross (two-factor cross)

trihybrid cross

forked-line (branch diagram) method

Stastical testing (analysis)

predicted occurrences

proportions

sample size

chance deviation

random fluctuations

null hypothesis

measured values
predicted values

goodness of fit

chi-square analysis (χ^2)

degrees of freedom

probability value (p)

reject the null hypothesis

fail to reject the null hypothesis

0.05 probability value

Pedigree

sibs, sibship line

monozygotic (identical) twins

dizygotic (fraternal) twins

proband

Concepts

unit factors in pairs (F3.1)

dominance/recessiveness (F3.1)

symbolism (F3.2)

segregation

phenotype

genotype

homozygous (homozygote)

heterozygous (heterozygote)

independent assortment

genotypic ratio

variation of a continuous nature

variation of a discontinuous nature

F 3.1. Illustration of the union of maternal and paternal genes (A and a) to give two genes in the zygote. Mendelian "unit factors" occur in pairs in diploid organisms. Dominant genes are often given the upper case letter as the symbol while the lower case letter is often used to symbolize the recessive gene.

It is important to see that each parent contributes one chromosome of each type (homologue) and thus one gene of each gene pair.

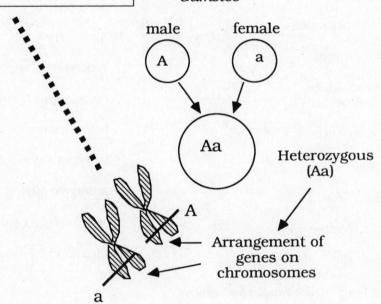

F3.2. It is important to see that in the drawing above there are four variations, but only two characteristics (shape, and color). When developing symbols for gene pairs, keep upper and lower case letters (or other compatible scheme) *of the same letter* to represent genes of the same characteristic.

Two characteristics but a total of four alternatives

Two characteristics
1. Shape (round, wrinkled)
2. Color (black, white)

Symbol choices (examples)

W = round   w = wrinkled
(assuming round is dominant
to wrinkled)

B = black   b = white
(assuming black is dominant
to white)

Selection of appropriate symbolism is critical

Solutions to Problems and Discussion Questions

1. Several points surface in the first sentence of this question. First, two alternatives (black and white) of one characteristic (coat color) are being described, therefore a monohybrid condition exists similar to K/C-Figs.3.1 and 3.2.

Second, are the guinea pigs in the parental generation (P_1) homozygous or heterozygous? Notice in the introductory sentence, just after PROBLEMS AND DISCUSSION QUESTIONS, there is the statement "members of the P_1 generation are homozygous..."

Third, which is dominant, *black* or *white*? Note that all the offspring are black, therefore black can be considered dominant. The second sentence of the problem verifies that a monohybrid cross is involved because of the 3/4 black and 1/4 white distribution in the offspring. Referring to K/C-Fig.3.2 and knowing that genes occur in pairs in diploid organisms, one can write the genotypes and the phenotypes requested in part (a) as follows:

(a)

P$_1$:

Phenotypes: Black   X   White

Genotypes: *WW*   *ww*

Gametes: (*W*)   (*w*)

F$_1$:   *Ww* (Black)

F$_1$ X   F$_1$:

Phenotypes: Black   X   Black

Genotypes: *Ww*   *Ww*

Gametes: (*W*) (*w*)   (*W*) (*w*)

(combine as in K/C-Fig.3.2)

F$_2$:

Phenotypes: Black   Black Black White

Genotypes: *WW*   *Ww* *Ww* *ww*

(b) *Since* white is a recessive gene (to *black*) each white guinea pig must be homozygous and a cross between two white guinea pigs must produce all white offspring.

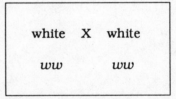

white   X   white

ww   *ww*

(c) Recall the various possibilities of the genotypes capable of producing the black phenotype in the F$_2$ generation in part (a) above: *WW* and *Ww*. In Cross 1 in the problem, all black offspring are observed and the most likely parental genotypes would be as follows:

WW   X   *WW*

or

WW   X   *Ww*

There is the possibility that black guinea pigs of the *Ww* genotype could produce all black offspring if the sample size was such that *ww* offspring were not produced. In Cross 2, a typical 3:1 Mendelian ratio is observed which indicates that two heterozygotes were crossed:

Ww X *Ww*.

See the F₁ cross in K/C-Figs.3.2 and 3.3.

2. Start out with the following gene symbols;

A =normal (not albino),

a=albino.

Since albinism is inherited as a recessive trait, genotypes *AA* and *Aa* should produce the normal phenotype, while *aa* will give albinism.

(a) The parents are both normal, therefore they could be either *AA* or *Aa*. The fact that they produce an albino child requires that each parent provides an *a* gene to the albino child; thus the parents must both be heterozygous(*Aa*).

See K/C-Fig.3.3.

(b) To start out, the normal male could have either the *AA* or *Aa* genotype. The female must be *aa*. Since all the children are normal one would consider the male to be *AA* instead of *Aa* . However, the male *could* be *Aa*. Under that circumstance, the likelihood of having six children, all normal is 1/64.

See K/C-Fig.3.4(a).

(c) To start out, the normal male could have either the *AA* or *Aa* genotype. The female must be *aa*. The fact that half of the children are normal and half are albino indicates a typical "test cross" in which the *Aa* male is mated to the *aa* female.

See K/C-Fig.3.4(b).

(d)

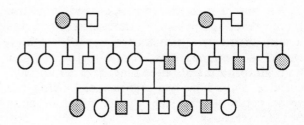

The 1:1 ratio of albino to normal in the last generation results because the mother is *Aa* and the father is *aa*.

3. *Unit Factors in Pairs*: It is important to see that each time a phenotype (normal or abnormal) is being stated, genotypes are symbolized as pairs of genes; *AA*, *Aa* or *aa*. Review F3.2 to understand the need to assign appropriate symbols to genes. Also see K/C-Fig.3.11(a).

Dominance and Recessiveness: Because the gene for normal pigmentation is completely dominant over the gene for albinism (*a* is fully recessive), it was necessary to consider, at first, whether normally pigmented individuals in the problem were homozygous normal (*AA*) or heterozygous (*Aa*). By looking at the frequency of expression of the recessive gene in the offspring (in *aa* individuals) one can often determine an *Aa* type from an *AA* type.

Segregation: During gamete formation when homologous chromosomes move to opposite poles, paired elements (genes) separate from each other.

See K/C-Fig.3.11(b).

4. While it is very difficult, if not impossible to know exactly how Mendel made the step from his "monohybrid results" to his postulates, he was able to develop a model with several important components or postulates.

First, organisms contained **unit factors** for various traits.

Second, if these **factors occurred in pairs**, there existed the possibility that some organisms would "breed true" if homozygous, while others would not (heterozygotes). If one "factor" of a pair had a **dominant influence** over the other, then he could explain how two organisms, looking the same, could be genetically different (homozygous or heterozygous).

Third, if the **paired elements separate (segregate)** from each other during gamete formation and if gametes combine at random, he could account for the 3:1 ratios in the monohybrid crosses.

Three excellent books give insight into Mendel's life and the context of his discoveries: Carlson, E. A. 1966. *The Gene: A Critical History*. Philadelphia: Saunders., Sturtevant, A. H. 1965. *A History of Genetics*. New York: Harper and Row. Voeller, B. R. 1968. *The Chromosome Theory of Inheritance*. New York: Appleton-Century-Crofts.

5. *Pisum sativum* is easy to cultivate. It is naturally self-fertilizing, but it can be crossbred. It has several visible features (e.g., tall or short, red flowers or white flowers) which are consistent under a variety of environmental conditions yet contrast due to genetic circumstances. Seeds could be obtained from local merchants.

6. With any "long" and involved problem, students often have trouble getting started in the right direction, and seeing the problem through to the necessary conclusions. First, read the entire question and see that you are to determine (1) the pattern of inheritance for "checkered and plain," and (2) the gene symbols and genotypes of all the parents and offspring. Notice that there is reference to one characteristic, *pattern*, with two alternatives, checkered vs. plain.

We should consider this to be a monohybrid condition unless complications arise. We are to use the respective offspring from the P_1 cross (F_1 progeny a, b, and c) in a series of F_1 crosses, d through g. Approach the problem by first assigning *probable* genotypes to the P_1 crosses then, where there is ambiguity (such as cross (a)), use the F_1 crosses for clarification.

Assignment of symbols:

P = checkered; p = plain. Checkered is tentatively assigned the dominant function because in a casual examination of the data, especially cross (b), we see that checkered types are more likely to be produced than plain types.

> Cross (a):
>
> *PP X PP*   or   *PP X Pp*

Notice in cross (d) that the checkered offspring, when crossed to plain, produce only checkered F_2 progeny and in cross (g) when crossed to checkered still produce only checkered progeny. From this additional information, one can conclude that in the progeny of cross (a) there are no heterozygotes and the original cross must have been *PP X PP*.

> Cross (b):
>
> *PP X pp*

This assignment seems reasonable because among 38 offspring, no plain types are produced. In addition, we would expect all the F_1 progeny to be heterozygous and if crossed to plain, as in cross (e), to produce approximately half checkered and half plain offspring. In cross (f) we would expect such heterozygotes to produce a 3:1 ratio, which is observed.

Cross (c):

Because all the offspring from this cross are plain, there is no doubt that the genotype of both parents is *pp*.

Genotypes of all individuals:

		F₁ Progeny	
P₁ Cross		*Checkered*	*Plain*
(a) *PP* X *PP*		*PP*	
(b) *PP* X *pp*		*Pp*	
(c) *pp* X *pp*			*pp*
(d) *PP* X *pp*		*Pp*	
(e) *Pp* X *pp*		*Pp*	*pp*
(f) *Pp* X *Pp*		*PP,Pp*	*pp*
(g) *PP* X *Pp*		*PP, Pp*	

7. In the first sentence you are told that there are two *characteristics* which are being studied; seed shape and cotyledon color. Expect, therefore, this to be a dihybrid situation with *two gene pairs* involved.   One also sees the possible alternatives of these two characteristics: *seed shape*; wrinkled vs. round; *cotyledon color*; green vs. yellow. After reading the second sentence you can predict that the gene for round seeds is dominant to that for wrinkled seeds and the gene for yellow cotyledons is dominant to the gene for green cotyledons.

Symbolism:

w = wrinkled seeds g = green cotyledons

W = round seeds G = yellow cotyledons

P₁:

   WWGG X wwgg

Parents are considered to be homozygous for two reasons. First, in the introductory sentence, just after PROBLEMS AND DISCUSSION QUESTIONS, there is the statement "members of the P₁ generation are homozygous..." Second, notice that the only offspring are those with round seeds and yellow cotyledons.

Gametes produced: One member of each gene pair is "segregated" to each gamete.

   WWGG   *wwgg*
   (WG)   (wg)

F₁: *WwGg*

F₁ X F₁:

   WwGg X WwGg

Gametes produced: Under conditions of independent assortment, there will be four (2^n, where n = number of heterozygous gene pairs) different types of gametes produced by each parent.

Punnett Square (See K/C-Fig.3.10):

	WG	*Wg*	*wG*	*wg*
WG	*WWGG*	*WWGg*	*WwGG*	*WwGg*
Wg	*WWGg*	*WWgg*	*WwGg*	*Wwgg*
wG	*WwGG*	*WwGg*	*wwGg*	*wwGg*
wg	*WwGg*	*Wwgg*	*wwGg*	*wwgg*

Collecting the phenotypes according the the dominance scheme presented above, gives the following:

9/16   W_G_   round seeds, yellow cotyledons

3/16   W_gg   round seeds, green cotyledons

3/16   wwG_   wrinkled seeds, yellow cotyledons

1/16   wwgg   wrinkled seeds, green cotyledons

Notice that a dash (_) is used where, because of dominance, it makes no difference as to the dominance/recessive status of the allele.

Forked, or branch diagram (See K/C-Fig.3.6):

Seed shape	Cotyledon color	Phenotypes
3/4 round	3/4 yellow	9/16 round, yellow
	1/4 green	3/16 round, green
1/4 wrinkled	3/4 yellow	3/16 wrinkled, yellow
	1/4 green	1/16 wrinkled, green

8. Symbolism as before:

w = wrinkled seeds	g = green cotyledons
W = round seeds	G = yellow cotyledons

Examine each characteristic (seed shape vs. cotyledon color) separately.

(a) Notice a 3:1 ratio for seed shape, therefore Ww X Ww; and no green cotyledons, therefore GG X GG or GG X Gg. Putting the two characteristics together gives

$$WwGG \quad X \quad WwGG$$

or

$$WwGG \quad X \quad WwGg.$$

(b) Notice a 1:1 ratio for seed shape (8/16 wrinkled and 8/16 round) and a 3:1 ratio for cotyledon color (12/16 yellow and 4/16 green). Therefore the answer is

$$wwGg \ X \ WwGg.$$

(c) The offspring occur in a typical 9:3:3:1 ratio, therefore the F_2 plants have the doubly heterozygous genotypes of

$$WwGg \ X \ WwGg.$$

(d) This is a typical 1:1:1:1 test cross (or backcross) ratio and will signify that one parent is doubly heterozygous while the other is fully homozygous recessive (see K/C-Fig.3.8). The answer is

$$WwGg \ X \ wwgg.$$

9. A test cross involves a fully heterozygous organism mated with a fully homozygous recessive organism. In Problem #8, (d) fits this description.

See K/C-Fig.3.8.

10. Because independent assortment may be defined as one gene pair segregating independently of another gene pair, one would need at least two gene pairs in order to demonstrate independent assortment.

11. Mendel's four postulates are related to this diagram.

1. Factors occur in pairs. Notice *A* and *a*.

2. Some genes are dominant to their alleles. Notice *A* and *a*.

3. Alleles segregate from each other during gamete formation. When homologous chromosomes separate from each other at anaphase I, alleles will go to opposite poles of the meiotic apparatus.

4. One gene pair separates independently from other gene pairs. Different gene pairs on the same homologous pair of chromosomes (if far apart) or on non-homologous chromosomes will separate independently from each other during meiosis.

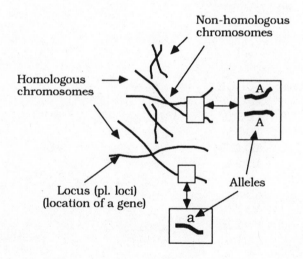

Non-homologous chromosomes

Homologous chromosomes

Locus (pl. loci) (location of a gene)

Alleles

12. Carefully re-read the answer to Question #2 in Chapter #2. Briefly, the factors which specify chromosomal homology are the following:

overall length,
position of the centromere,
banding patterns,
type and location of genes,
autoradiographic pattern.

13. Homozygosity refers to a condition where both genes of a pair are the same (i.e. *AA* or *GG* or *hh*) whereas heterozygosity refers to the condition where members of a gene pair are different (i.e. *Aa* or *Gg* or *Bb*). Homozygotes produce only one type of gamete whereas heterozygotes will produce 2^n types of gametes where n = number of heterozygous gene pairs (assuming independent assortment).

14. There are two characteristics presented here, body color and wing length. First, assign meaningful gene symbols.

	Body color	*Wing length*
	E = grey body color	*V* = long wings
	e = ebony body color	*v* = vestigial wings

(a)

P_1:

EEVV   X   eevv   Refer to K/C-Fig.3.7

F_1: *EeVv* (grey, long)

F_2:

This will be the result of a Punnett Square with 16 boxes as in K/C-Fig.3.7.

Phenotypes Ratio   Genotypes Ratio

grey , long	9/16	*EEVV*	1/16
		EEVv	2/16
		EeVV	2/16
		EeVv	4/16
grey, vestigial	3/16	*EEvv*	1/16
		Eevv	2/16
ebony, long	3/16	*eeVV*	1/16
		eeVv	2/16
ebony, vestigial	3/16	*eevv*	1/16

(b)

P₁:

> *EEvv X eeVV*

F₁: It is important to see that the results from this cross will be exactly the same as those in part (a) above. The only difference is that the recessive genes are coming from both parents, rather than from one parent only as in (a). The F₂ ratio will be the same as (a) also. When you have genes on the autosomes (not X-linked) independent assortment, complete dominance, and no gene interaction (see later) in a cross involving double heterozygotes, the offspring ratio will be in the ratio 9:3:3:1.

(c)

P₁:

> *EEVV X EEvv*

F₁: *EEVv* (grey, long)

F₂: Notice that all the offspring will have grey bodies and you will get a 3:1 ratio of long to vestigial wings. You should see this before you even begin working through the problem. Even though this cross involves two gene pairs it will give a "monohybrid " type of ratio because one of the gene pairs is homozygous (body color) and **one** gene pair is heterozygous (wing length).

Phenotypes	Ratio	Genotypes	Ratio
grey, long	3/4	*EEVV*	1/4
		EEVv	2/4
grey, vestigial	1/4	*EEvv*	1/4

NOTE: After working through this problem, it is important that you try to work similar problems without constructing the time-consuming Punnett squares especially if each problem asks for phenotypic rather than genotypic ratios.

15. The general formula for determining the number of kinds of gametes produced by an organism is 2^n where n = number of *heterozygous* gene pairs.

(a) 4: AB, Ab, aB, ab

(b) 2: AB, aB

(c) 8: ABC, ABc, AbC, Abc, aBC, aBc, abC, abc

(d) 2: ABc, aBc

(e) 4: ABc, Abc, aBc, abc

(f) $2^5 = 32$

ABCDE	aBCDE
ABCDe	aBCDe
ABCdE	aBCdE
ABCde	aBCde
ABcDE	aBcDE
ABcDe	aBcDe
ABcdE	aBcdE
ABcde	aBcde
AbCDE	abCDE
AbCDe	abCDe
AbCdE	abCdE
AbCde	abCde
AbcDE	abcDE
AbcDe	abcDe
AbcdE	abcdE
Abcde	abcde

Notice that there is a pattern that can be used to write these gametes so that fewer errors will occur.

16.

(a) When examining this cross

AaBbCc X AaBBCC

expect there to be eight different kinds of gametes from one parent (*AaBbCc*), and two different kinds from the other (AaBBCC). Therefore there should be sixteen kinds (genotypes) of offpsring (8 X 2).

Gametes: Gametes:

ABC		ABC
ABc		aBC
AbC		
Abc		
aBC		
aBc		
abC		
abc		

Offspring:

Genotypes	Ratio	Phenotypes
AABBCC	(1/16)	
AABBCc	(1/16)	
AABbCC	(1/16)	
AABbCc	(1/16)	
AaBBCC	(2/16)	A_B_C_ = 12/16
AaBBCc	(2/16)	
AaBbCC	(2/16)	
AaBbCc	(2/16)	
aaBBCC	(1/16)	
aaBBCc	(1/16)	aaB_C_ = 4/16
aaBbCC	(1/16)	
aaBbCc	(1/16)	

(b) There will be four kinds of gametes for the first parent (AaBBCc) and two kinds of gametes for the second parent.

Gametes: Gametes:

ABC		aBC
ABc		aBc
aBC		
aBc		

Offspring:

Genotypes	Ratio	Phenotypes	
AaBBCC	1/8	A_BBC_	= 3/8
AaBBCc	2/8		
AaBBcc	1/8	A_BBcc	= 1/8
aaBBCc	1/8	aaBBC_	= 3/8
aaBBCC	2/8		
aaBBcc	1/8	aaBBcc	= 1/8

(c) There will be eight (2^n) different kinds of gametes from each of the parents, therefore a 64-box Punnett square. Doing this problem by the forked-line method helps considerably.

```
                                  /  1/4 CC = 1/64 AABBCC
                        1/4 BB  — 2/4 Cc  = 2/64 AABBCc
                       /          \ 1/4 cc
                      /           /  1/4 CC        etc.
          1/4 AA — 2/4 Bb  — 2/4 Cc
                      \           \ 1/4 cc
                       \          /  1/4 CC
                        1/4 bb  — 2/4 Cc
                                  \ 1/4 cc
```

		1/4 CC
	1/4 BB	2/4 Cc
		1/4 cc
		1/4 CC
2/4 Aa	2/4 Bb	2/4 Cc
		1/4 cc
		1/4 CC
	1/4 bb	2/4 Cc
		1/4 cc
		1/4 CC
	1/4 BB	2/4 Cc
		1/4 cc
		1/4 CC
1/4 aa	2/4 Bb	2/4 Cc
		1/4 cc
		1/4 CC
	1/4 bb	2/4 Cc
		1/4 cc

Simply multiply through each component to arrive at the final genotypic frequencies.

For the phenotypic frequencies, set up the problem in the following manner (as in K/C-Fig.3.10).

$$
3/4 \text{ A\_} \begin{cases} 3/4 \text{ B\_} \begin{cases} 3/4 \text{ C\_} = 27/64 \text{ A\_B\_C\_} \\ 1/4 \text{ cc} = 9/64 \text{ A\_B\_cc} \end{cases} \\ 1/4 \text{ bb} \begin{cases} 3/4 \text{ C\_} \quad \text{etc.} \\ 1/4 \text{ cc} \end{cases} \end{cases}
$$

$$
1/4 \text{ aa} \begin{cases} 3/4 \text{ B\_} \begin{cases} 3/4 \text{ C\_} \\ 1/4 \text{ cc} \end{cases} \\ 1/4 \text{ bb} \begin{cases} 3/4 \text{ C\_} \\ 1/4 \text{ cc} \end{cases} \end{cases}
$$

17. In the first reading of this question one should consider that there is one characteristic involved, seed color. Given that information and the F_2 progeny of 6022 yellow and 2001 green, a monohybrid condition is suspected. In addition, of the 519 self-fertilized, yellow seeded plants, 166 bred true while the others produced a 3:1 ratio of yellow to green. This is what would be expected because approximately 1/3 of the yellow F_2 plants should be homozygous, while 2/3 of the yellow F_2 plants should be heterozygous.

Set up the symbols as follows: G = yellow seeds, g = green seeds. We know that the gene for yellow seeds is dominant to that for gren seeds because the F_1 is all yellow, not green.

	Phenotypes	*Genotypes*
P_1:	Yellow X green	GG X gg
F_1:	all yellow	Gg
F_2:	6022 yellow	1/4 GG; 2/4 Gg
	2001 green	1/4 gg

Of the yellow F_2 offspring, notice that 1/3 of them are GG and 2/3 are Gg. If you selfed the 1/3 GG types then all the offspring (the 166) would breed true whereas the others (353 which are Gg) should produce offspring in a 3:1 ratio when selfed.

GG X GG
$= $ all GG
Gg X Gg
$= 1/4\ GG;\ 2/4\ Gg;\ 1/4\ gg$

18. Because there is only one characteristic being dealt with in this problem (coat color) and a 3/1 ratio is mentioned in the second sentence, one can initially consider this to be a monohybrid condition. Set the gene symbols as W = black, w = white.

One hundred black animals could be all WW, all Ww, or a mixture (WW, Ww). When the WW individuals are crossed to ww, all the offspring will be black, which is what occurred in 94 of the cases. All these black offspring would be heterozygous and be expected to produce a 3:1 ratio when intercrossed. In those cases where a 1:1 ratio resulted, the parents must have been Ww. The cross would be as follows:

P_1:
Ww X ww
F_1:
1/2 Ww, 1/2 ww

If one were to cross the black and white guinea pigs from the above cross, again the offspring would produce 1/2 black and 1/2 white as above.

19. In reading this question, notice that there are two characteristics being considered; seed color (yellow, green) and seed shape (round, wrinkled). At this point you should be able to do this problem without writing down each of the steps. The F_1 can be considered to be a double heterozygote (with round and yellow being dominant). See the cross this way:

Symbols:

Seed shape	Seed color
W = round	G = yellow
w = wrinkled	g = green

P_1:

WWgg X wwGG

F_1:   WwGg cross to wwgg

(which is a typical test cross)

The offspring will occur in a typical 1:1:1:1 as

1/4 WwGg (round, yellow)

1/4 Wwgg (round, green)

1/4 wwGg (wrinkled, yellow)

1/4 wwgg (wrinkled, green)

Again, at this point it would be very helpful if you could do such simple problems by inspection.

20. This question deals with the definition of dominance/recessiveness. Notice that there are only two alleles (one gene pair) and three phenotypes associated with the problem; normal, "minor" anemia and "major" anemia. Under a monohybrid model, the heterozygote is distinguishable from either homozygote which does not fit the definition of dominance. One would conlcude that no dominance is involved and incomplete dominance exists (see Chapter 4).

21. Since these are F_2 results from monohybrid crosses a 3:1 ratio is expected for each. Referring to K/C-Table3.1 on can set up the analysis easily.

(a)

Expected ratio	Observed (o)	Expected (e)
3/4	882	885.75
1/4	299	295.25

Expected values are derived by multiplying the expected ratio by the total number of organisms.

$$\chi 2 = \Sigma \frac{(o - e)2}{e} = .064$$

In looking in the χ^2 table (K/C-Fig.3.12) with 1 degree of freedom (because there were two classes, therefore n-1 or 1 degree of freedom, we find a probability (P) value of between 0.9 and 0.5.

We would therefore say that there is a "good fit" between the observed and expected values. Notice that as the deviations between the observed and expected values increase the value of χ^2 increases. So the higher the χ^2 value the more likely the null hypothesis will be rejected.

(b)

Expected ratio	Observed (o)	Expected (e)
3/4	705	696.75
1/4	224	232.25

$$\chi^2 = 0.39$$

The *P* value in the table for 1 degree of freedom is still between 0.9 and 0.5, however because the χ^2 value is larger in (b) we sould say that the deviations from expectation are greater.

22. One must think of this problem as a dihybrid F_2 situation with the following expectations:

Expected ratio	Observed (o)	Expected (e)
9/16	315	312.75
3/16	108	104.25
3/16	101	104.25
1/16	32	34.75

$$\chi^2 = 0.5$$

Looking at the table in K/C-Fig.3.12 one can see that this χ^2 value is associated with a probability greater than 0.90 for 3 degrees of freedom (because there are now four classes in the χ^2 test). The observed and expected values do not deviate significantly.

To deal with parts **(b)** and **(c)** it is easier to see the observed values for the monohybrid ratios if the phenotypes are listed:

smooth, yellow	315
smooth, green	108
wrinkled, yellow	101
wrinkled, green	32

For the smooth: wrinkled *monohybrid component*, the smooth types total 423 (315 + 108), while the wrinkled types total 133 (101 + 32).

Expected ratio	Observed (o)	Expected (e)
3/4	423	417
1/4	133	139

The χ^2 value is 0.35 and in examining K/C-Fig.3.12 for 1 degree of freedom, the P value is greater than 0.50 and less than 0.90. We fail to reject the null hypothesis and are confident that the observed values do not differ significantly from the expected values.

(c) For the yellow:green portion of the problem, see that there are 416 yellow plants (315 + 101) and 140 (108 + 32) green plants.

Expected ratio	Observed (o)	Expected (e)
3/4	416	417
1/4	140	139

The χ^2 value is 0.01 and in examining K/C-Fig.3.12 for 1 degree of freedom, the P value is greater than 0.90. We fail to reject the null hypothesis and are confident that the observed values do not differ significantly from the expected values.

23. It would be best to set up two tables based on the two hypotheses:

Expected ratio	Observed (o)	Expected (e)
3/4	250	300
1/4	150	100

Expected ratio	Observed (o)	Expected (e)
1/2	250	200
1/2	150	200

For the test of a 3:1 ratio, the χ^2 value is 33.3 with an associated P value of less than 0.01 for 1 degree of freedom.

For the test of a 1:1 ratio, the χ^2 value is 25.0 again with an associated P value of less than 0.01 for 1 degree of freedom. Based on these probability values, both null hypotheses should be rejected.

24. Use of the P = 0.10 as the "critical" value for rejecting or failing to reject the null hypothesis instead of P = 0.05 would allow more null hypotheses to be rejected. Notice in K/C-Fig.3.12 that as the χ^2 values increase, there is a higher likelihood that the null hypothesis will be rejected because the higher values are more likely to be associated with a P value which is less than 0.05.

As the critical P value is increased, it takes a smaller χ^2 value to cause rejection of the null hypothesis. It would take less difference between the expected and observed values to reject the null hypothesis, therefore the stringency of failing to reject the null hypothesis is increased.

25. While there are many different inheritance patterns which will be described later in the K/C text (codominance, incomplete dominance, sex-linked inheritance, etc.) the range of solutions to this question is limited to the concepts developed in the first three chapters, namely dominance or recessiveness.

If a gene is dominant, it will not skip generations nor will it be passed to offspring unless the parents possess the gene. On the other hand, genes which are recessive can skip generations and exist in a carrier state in parents. For example, notice that II-4 and II-5 produce a female child (III-4) with the affected phenotype. On these criteria alone, the gene must be viewed as being recessive. Note: if a gene is recessive and X-linked (to be discussed later) the pattern will often be from affected male to carrier female to affected male.

To provide genotypes for each individual consider that if the box or circle is shaded, the *aa* genotype is to be assigned. If offspring are affected (shaded) a recessive gene must have come from both parents.

I-1 (*aa*), I-2 (*Aa*), I-3 (*Aa*), I-4 (*Aa*)

II-1 (*aa*), II-2 (*Aa*), II-3 (*aa*), II-4 (*Aa*), II-5 (*Aa*), II-6 (*aa*), II-7 (*AA* or *Aa*), II-8 (*AA* or *Aa*)

III-1 (*AA* or *Aa*), III-2 (*AA* or *Aa*), III-3 (*AA* or *Aa*), III-4 (*aa*), III-5 (probably *AA*), III-6 (*aa*)

IV-1 through IV-7 all *Aa*.

26. Applying the same logic as in question #25, the gene is inherited as an autosomal recessive. Notice that two normal individuals II-3 and II-4 have produced a daughter (III-2) with myopia.

I-1 (*AA* or *Aa*), I-2 (*aa*), I-3 (*Aa*), I-4 (*Aa*)

II-1 (*Aa*), II-2 (*Aa*), II-3 (*Aa*), II-4 (*Aa*), II-5 (*aa*), II-6 (*AA* or *Aa*), II-7 (*AA* or *Aa*)

III-1 (*AA* or *Aa*), III-2 (*aa*), III-3 (*AA* or *Aa*)

27. There are two possibilities. Either the trait is dominant, in which case I-2 is heterozygous as are II-2 and II-3, or the trait is recessive and I-1 is heterozygous and I-2 is homozygous recessive. Under the condition of recessiveness, II-1 and II-4 would be heterozygous, II-2 and II-3 homozygous.

28. Given the cross *AaBbCC X AABbCc* we can apply the product rule which states that when two or more events occur independently but simultaneously, their combined probability is equal to the product of their individual probabilities.

> The probability of getting *AA* from
>
> $\quad$ *Aa   X   AA* is 1/2
>
> The probability of getting *Bb* from
>
> $\quad$ *Bb   X   Bb* is 1/2
>
> The probability of getting *Cc* from
>
> $\quad$ *CC   X   Cc* is 1/2
>
> The overall probability then is
>
> $\quad$ 1/2 X 1/2 X 1/2 = 1/8

29. The probability of getting *aabbcc* from the *AaBbCC X AABbCc* mating is zero because of homozygosity for *AA* and *CC*.

30. Because all the offspring will show the dominant A and C phenotypes and 3/4 will show the B phenotype, the probability of an offspring showing all three dominant traits would be

$$1 \ X \ 3/4 \ X \ 1 \ = \ 3/4.$$

4

Modification of Mendelian Ratios

Vocabulary: Organization and Listing of Terms

Structures and Substances

Native antigens

Antiserum

Antibodies

Ommatidia

   drosopterin

   xanthommatin

Processes/Methods

Incomplete (partial) dominance

   pink flowers

   1:2:1 phenotypic ratio

Codominance

   MN blood groups

Multiple allelism

   ABO blood types

antigen-antibody reaction

   cross-reaction

   agglutination

   isoagglutinogen

   H substance

   Bombay phenotype

   secretor locus

   Rh antigens

   erythroblastosis fetalis (HDN)

white eye in *Drosophila*

Lethal alleles

   recessive

   dominant

     yellow coat color in mice

     Huntington's disease

Gene interaction: discontinuous variation

epistasis

    homozygous recessive

        coat color in mice

        Bombay phenotype

        9:3:4

    dominant

        fruit color in squash

        12:3:1

    other

        white flowers in peas

        9:7

    novel phenotypes

        fruit shape in *Cucurbita*

        eye color in *Drosophila*

Gene interaction: continuous variation

    polygenic

    multiple-factor (multi-gene)

    additive effects

        1:4:6:4:1

        $2n+1$

Concepts

Gene interaction (F4.1)

Neo-Mendelian genetics

Allele (F4.2)

    wild type

    mutation

        loss of wild type function

        reduced or increased function

Symbolism (F4.2)

    recessive trait (e, e^+)

    dominant trait (Wr, Wr^+)

    "+" as a superscript

    + as a symbol for wild type ($+/e$)

    no dominance (R1, R2)

Modified ratios (T4.1)

    3:1

        1:2:1

    9:3:3:1

        3:6:3:1:2:1

        9:3:4

        12:3:1

        9:7

        1:4:6:4:1

T4.1. Examples of typical monohybrid and dihybrid ratios and several modifications.

Basic Ratio	Modification	Explanation
3:1	1:2:1	Incomplete dominance
		Codominance
9:3:3:1	9:3:4	Epistasis
	12:3:1	Epistasis
	9:7	Epistasis
	3:6:3:1:2:1	Dominance + incomplete dominance or codominance
	1:4:6:4:1	Additive effects

F4.1. Illustration of gene interaction where products from more than one gene pair influence one characteristic or phenotypic trait.

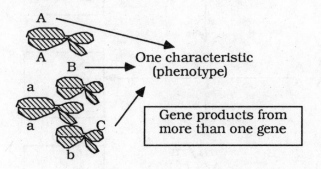

F4.2. Illustration of the symbolism associated with the wild type activity of a gene and several possible outcomes of the mutant state (m/m).

First, too much product (B) is made in the mutant state.

Second, too little product is made.

Third, no product is made.

The presence of the wild type allele (+) allows for the normal conversion
        of substrate A to product B.

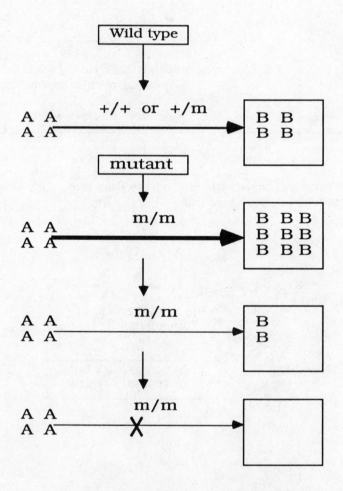

Solutions to Problems and Discussion Questions

1. In the first sentence of this problem, notice that there is one characteristic (coat color) and three phenotypes mentioned; red, white, or roan. The fact that roan is intermediate between red and white suggests that this may be a case of incomplete dominance, with roan being the intermediate and therefore the heterozygous type as is "pink" in K/C-Fig.4.1. If that is the case then we should suspect a 1:2:1 phenotypic ratio in crosses of "roan to roan."

Looking at the data given, notice that a cross of the "extremes" (red X white) gives roan, suggesting its heterozygous nature and the homozygous nature of the parents. Seeing the 1:2:1 ratio in the offspring of

roan X roan

confirms the hypothesis of incomplete dominance as the mode of inheritance.

Symbolism:

 AA = red,

 aa = white,

 Aa = roan

Crosses: It is important at this point that you not be fully dependent on writing out complete Punnett squares for each cross. Begin working these simple problems in your head.

AA	X	AA	—>	AA
aa	X	aa	—>	aa
AA	X	aa	—>	Aa
Aa	X	Aa	—>	1/4 AA;
				2/4 Aa; 1/4 aa

2. *Incomplete dominance* can be viewed more as a quantitative phenomenon where the heterozygote is intermediate (approximately) between the limits set by the homozygotes. Pink is intermediate between red and white.

Codominance can be viewed in a more qualitative manner where both of the alleles in the heterozygote are expressed. For example in the AB blood group, both the I^A and I^B genes are expressed . There is no intermediate class which is part I^A and I^B.

3. Notice that there is one typical (coat color) and one atypical (lethality) characteristic mentioned. Often under this condition of two characteristics, we must decide if the problem involves one or more than one gene pair. Because the genotypes are given here it is obvious that lethality is associated with expression of the coat color alleles and therefore one gene pair is involved. This is a monohybrid condition.

Pp X Pp —>

 1/4 PP (**lethal**)

 2/4 Pp (platinum)

 1/4 pp (silver)

Therefore the ratio of surviving foxes is 2/3 platinum, 1/3 silver. The P allele behaves as a recessive in terms of lethality (seen only in the homozygote) but as a dominant in terms of coat color (seen in the homozygote). This cross is similar to Cross B in K/C-Fig.4.4.

4. In this problem it would be helpful to first diagram the phenotypes of the crosses so that you can get some idea of the the inheritance pattern. From those phenotypic crosses, a suggestion as to the genotypes can be made and verified.

Cross 1:

short tail   X   normal long tail

approximately 1/2 short, 1/2 long

This tells you that one type is heterozygous, the other homozygous.

Cross 2:

short tail   X   short tail

6 short tail, 3 long tail (2/3 short, 1/3 long)

At this point one would consider that the 2/3 short are heterozygotes, and the long is the homozygous class. Also short is dominant to long. Since these ratios were repeated and verified, one can conclude that a 2:1 ratio is not a statistical artifact and that the following genotypic model would hold. Because long is the "normal" does not mean that it is dominant.

Symbolism:

S = short, s = long

Cross 1:

$S s$ X ss —>

1/2 Ss (short), 1/2 ss (long)

Cross 2:

Ss X Ss —>

1/4 SS (lethal), 2/4 Ss (short), 1/4 ss (long)

See K/C-Fig.4.4, especially Cross B and Cross C, for another example of this type of cross.

5. In the section on multiple alleles in the K/C text, there is a table which indicates all the genotypes requested.

Blood Group (phenotype)	Genotype(s)
A	$I^A I^A$, $I^A I^o$
B	$I^B I^B$, $I^B I^o$
AB	$I^A I^B$
O	$I^o I^o$

I^A and I^B *are* codominant (notice the AB blood group) while being dominant to I^o.

6. In this problem remember that individuals with blood type B can have the genotype $I^B I^B$ or $I^B I^o$ and those with blood type A, genotypes $I^A I^A$ or $I^A I^o$.

Male Parent: must be $I^B I^o$ because the mother is $I^o I^o$ and one inherits one homologue (therefore one allele) from each parent.

Female Parent: must be $I^A I^o$ because the father is $I^B I^o$ and one inherits one homologue (therefore one allele) from each parent. The father can not be $I^B I^B$ and have a daughter of blood type A.

Offspring:

$I^A I^o$ X $I^B I^o$

	I^B	I^o
I^A	$I^A I^B$(AB)	$I^A I^o$(A)
I^o	$I^B I^o$ (B)	$I^o I^o$(O)

The ratio would be 1:1:1:1.

7. Given that a child is blood type O (genotype $I^o I^o$) and the mother blood type A (she must be $I^A I^o$ to have had a type O child), then the father could have the following genotypes: $I^B I^o$, $I^A I^o$ or $I^o I^o$. In other words, the father must have been able to contribute I^o to the child.

The only *blood type* which would exclude a male from being the father would be AB, because no I^o allele is present. Because many individuals in a population could have genotypes with the I^o allele, one could not prove that a particular male was the father by this method.

8. First, as before, notice that one characteristic with two alternatives (secretor, nonsecretor) is being discussed. By looking at the "mixed" cross (secretor X nonsecretor), one can determine which gene is dominant. That the secretor gene is dominant is supported by the cross involving F_1 secretors which gives

3/4 secretors:1/4 nonsecretors.

From this information it seems reasonable to assign a dominant function to the secretor allele.

Symbolism:

S = secretor s = nonsecretor

Cross 1:

SS X SS —> SS

Cross 2:

ss X ss —> ss

Cross 3: SS X ss —> Ss

(the secretor parent is known to be homozygous because of all secretor ofspring)

Cross 4:

Ss X Ss —> 1/4 SS; 2/4 Ss; 1/4 ss

9. Given that there are 100 amino acids and each position can theoretically have 20 different amino acids, there would be 100^{20} different alleles possible. In practice, it is likely that some of the substitutions would not drastically influence the phenotype of the organism. This subject will be discussed in later chapters.

10. It is important to see that this problem involves multiple alleles, meaning that monohybrid type ratios are expected, and that there is an order of dominance which will allow certain alleles to be "hidden" in various heterozygotes. As with most genetics problems, one must look at the phenotypes of the offspring to assess the genotypes of the parents.

(a)

Phenotypes:

   Himalayan X Himalayan —> albino

Genotypes: $c^h c^a$ $c^h c^a$ $c^a c^a$

The Himalayan parents must both be heterozygous to produce an albino offspring.

Phenotypes:

   full color X albino —> chinchilla
Genotypes: Cc^{ch} $c^a c^a$ $c^{ch} c^a$

Because of the cc albino parent, the genotype of the chinchilla F_1 must be $c^{ch} c^a$. Also in order to have a chinchilla offspring at all, the full color parent must he heterozygous for chinchilla.

Therefore the cross of albino with chinchilla would be as follows:

$c^a c^a$ X $c^{ch} c^a$ —>

   1/2 chinchilla; 1/2 albino

(b)

Phenotypes:

   albino X chinchilla —> albino

Genotypes: $c^a c^a$ $c^{ch} c^a$ $c^a c^a$

Phenotypes:

   full color X albino —> full color
Genotypes: $C\_$ $c^a c^a$ Cc^a

It is impossible to determine the complete genotype of the full color parent but the full color offspring must be as indicated, Cc^a.

Therefore the cross of the albino with full color would be as follows:

$c^a c^a$ X Cc^a —>

   1/2 full color; 1/2 albino

(c)

Phenotypes:

   chinchilla X albino —> Himalayan

Genotypes: $c^{ch} c^h$ $c^a c^a$ $c^h c^a$

The chinchilla parent must be heterozygous for Himalayan because of the Himalayan offspring.

Phenotypes:

   full color X albino —> Himalayan

Genotypes: Cc^h $c^a c^a$ $c^h c^a$

Therefore a cross between the two Himalayan types would produce the following offpsring:

$c^h c^a$ X $c^h c^a$ —>

   3/4 Himalayan; 1/4 albino

11. It is important to see that this problem involves multiple alleles, meaning that monohybrid type ratios are expected, and that there is an order of dominance which will allow certain alleles to be "hidden" in various heterozygotes. As with most genetics problems, one must look at the phenotypes of the offspring to assess the genotypes of the parents.

(a)

Parents: sepia X cream

Because both guinea pigs had albino parents, both are heterozygous for the c^a allele.

Cross:

$c^k c^a$ X $c^d c^a$ —>

2/4 sepia; 1/4 cream; 1/4 albino

(b)

Parents: sepia X cream

Because the sepia parent had an albino parent it must be $c^k c^a$. Because the cream guinea pig had two sepia parents

($c^k c^d$ X $c^k c^d$ or $c^k c^d$ X $c^k c^a$),

the cream parent could be $c^d c^d$ or $c^d c^a$.

Crosses:

$c^k c^a$ X $c^d c^d$ —>

1/2 sepia; 1/2 cream*

*(if parents are assumed to be homozygous)

or

$c^k c^a$ X $c^d c^a$ —>

1/2 sepia; 1/4 cream; 1/4 albino

(c)

Parents: sepia X cream

Because the sepia guinea pig had two full color parents which could be

Cc^k, Cc^d, or Cc^a

(not CC because sepia could not be produced),

its genotype could be

$c^k c^k$, $c^k c^d$, or $c^k c^a$.

Because the cream guinea pig had two sepia parents

($c^k c^d$ X $c^k c^d$ or $c^k c^d$ X $c^k c^a$),

the cream parent could be $c^d c^d$ or $c^d c^a$.

Crosses:

$c^k c^k$ X $c^d c^d$ —> all sepia

$c^k c^k$ X $c^d c^a$—> all sepia

$c^k c^d$ X $c^d c^d$ —> 1/2 sepia; 1/2 cream

$c^k c^d$ X $c^d c^a$ —> 1/2 sepia; 1/2 cream

$c^k c^a$ X $c^d c^d$ —> 1/2 sepia; 1/2 cream

$c^k c^a$ X $c^d c^a$ —>

　　　　1/2 sepia; 1/4 cream; 1/4 albino

(d)

Parents: sepia　X　cream

Because the sepia parent had a full color parent and an albino parent

$$(Cc^k \quad X \quad c^a c^a),$$

it must be $c^k c^a$. The cream parent had two full color parents which could be Cc^d or Cc^a; therefore it could be $c^d c^d$ or $c^d c^a$.

Crosses:

$c^k c^a$ X $c^d c^d$ —> 1/2 sepia; 1/2 cream

$c^k c^a$ X $c^d c^a$ —>

　　　　1/2 sepia; 1/4 cream; 1/4 albino

12. Three independently assorting characteristics are being dealt with: flower color (incomplete dominance), flower shape (dominant/recessive), and plant height (dominant/recessive). Establish appropriate gene symbols:

Flower color: (see K/C-Fig.4.1)

　　RR = red ; Rr = pink; rr = white

Flower shape:

　　P = personate; p = peloric

Plant height:

　　D = tall; d = dwarf

(a)

$RRPPDD$　X　$rrppdd$　—>

　　$RrPpDd$ (pink, personate, tall)

(b) Use *components* of the forked line method as follows:

2/4 pink　X　3/4 personate　X　3/4 tall

　　　　= 18/64

13. There are two characteristics, flower color and flower shape. Because pink results from a cross of red and white, one would conclude that flower color is "monohybrid" with incomplete dominance.

In addition, because personate is seen in the F_1 when personate and peloric are crossed, personate must be dominant to peloric. Results from crosses (c) and (d) verify these conclusions. The appropriate symbols would be as follows (See K/C-Fig.4.1):

Flower color:

   RR = red; Rr = pink; rr = white

Flower shape:

   P = personate; p = peloric

(a)

   $RRpp$ X $rrPP$ ---> $RrPp$

(b)

   $RRPP$ X $rrpp$ ---> $RrPp$

(c)

   $RrPp$ X $RRpp$ ---> $RRPp$
   $RRpp$
   $RrPp$
   $Rrpp$

(d)

   $RrPp$ X $rrpp$ ---> $rrPp$
   $rrpp$
   $RrPp$
   $Rrpp$

In the cross of the F_1 of (a) to the F_1 of (b), both of which are double heterozygotes, one would expect the following:

$$RrPp \ X \ RrPp$$

1/4 red ⟨ 3/4 personate —> 3/16 red, personate
          1/4 peloric   —> 1/16 red, peloric

2/4 pink ⟨ 3/4 personate —> 6/16 pink, personate
           1/4 peloric   —> 2/16 pink, peloric

1/4 white ⟨ 3/4 personate —> 3/16 white, personate
            1/4 peloric   —> 1/16 white, peloric

14. This is a case of gene interaction (novel phenotypes) in which the recessive, independently assorting genes *brown* and *scarlet* (both recessive) interact to give the white phenotype. Refer to K/C-Fig.4.9 and see that the symbolism uses a "+" superscript to indicate the wild type. For simplicity in this problem, assume that all parental crosses involve homozygotes.

(a)

$bw^+/bw^+;st^+/st^+$ X $bw/bw;st/st$ —>

   bw^+/bw; st^+/st (wild)

The F_2 would produce the expected 9:3:3:1 ratio except that gene interaction will give the white phenotype in the 1/16 class.

$bw^+/\_$; $st^+/\_$	= wild type
$bw^+/\_$; st/st	= scarlet
bw /bw ; $st^+/\_$	= brown
bw /bw ; st /st	= white

(b)

$$bw^+/bw^+; st^+/st^+ \quad X \quad bw^+/bw^+; st/st \longrightarrow$$

$$bw^+/bw^+; \ st^+/st \ \text{(wild)}$$

The F_2, resulting from a cross of

$$bw^+/bw^+; \ st^+/st \quad X \quad bw^+/bw^+; \ st^+/st$$

would produce a 3:1 ratio of wild to scarlet.

(c)

$$bw/bw; st^+/st^+ \quad X \quad bw/bw; st/st \longrightarrow$$

$$bw/bw; \ st^+/st \ \text{(brown)}$$

The F_2, resulting from a cross of

$$bw/bw; \ st^+/st \quad X \quad bw/bw; \ st^+/st$$

would produce a 3:1 ratio of brown to white. Notice in the F_2 crosses in parts (b) and (c) one of the parents in each is homozygous, therefore the crosses will give monohybrid types of ratios.

15. This is a case in which epistasis (from *cc*) results in a "masking" of genes at the *A* locus (see case #1 in K/C-Fig.4.8). In this case there will be modifications of typical 9:3:3:1 and 1:1:1:1 ratios because of gene interactions.

(a) In a cross of

$$AACC \ X \ aacc,$$

the offpsring are all *AaCc* (gray) because the *C* allele allows pigment to be deposited in the hair and when it is it will be gray.

F_2 offspring would have the following "simplified" genotypes with the corresponding phenotypes:

$A\_C\_ \ = 9/16$ (gray)

$A\_cc \ = 3/16$
   (colorless because *cc* is epistatic to *A*)

$aaC\_ \ = 3/16$ (black)

$aacc \ = 1/16$
   (colorless because *cc* is epistatic to *aa*)

The two colorless classes are phenotyically indistinguishable, therefore the final ratio is 9:3:4.

(b) Results of crosses of female gray

$$(A\_C\_) \ X \ aacc \ \text{(males)}$$

are given in three groups:

(1) To produce an even number of gray and colorless offspring, the female parent must have been *AACc* so that half of the offspring are able to deposit pigment because of *C* and when they do, they are all gray (having received only *A* from the female parent).

(2) To produce an even number of gray and black offspring the mother must have been *Aa* and so that no colorless offpring were produced the female must have been *CC*. Her genotype must have been *AaCC*.

(3) Notice that half of the offspring are colorless, therefore the female must have been *Cc*. Half of the pigmented offspring are black and half are gray, therefore the female must have been *Aa*. Overall, the *AaCc* genotype seems appropriate.

16. Notice that the distribution of observed offspring fits a 9:3:4 ratio quite well. This suggests that two independently assorting gene pairs with epistasis are involved. Assign gene symbols in the usual manner:

> *A* = pigment; *a* = pigmentless (colorless)
>
> *B* = purple; *b* = red

> *AaBb* X *AaBb*
>
>   produces
>
> *A_B_* = purple
> *A_bb* = red
> *aaB_* = colorless
> *aabb* = colorless

One may see this occurring in the following manner:

> precursor —+-> cyanidin —+-> purple pigment
> (colorless) *aa*   (red)   *bb*

17. This is a case of gene interaction (novel phenotypes) where the yellow and black types (double mutants) interact to give the cream phenotype and epistasis where the *cc* genotype produces albino.

(a)

> *AaBbCc* —>
>
>   gray (*C* allows pigment)

(b)

> *A_B_Cc* —>
>
>   gray (*C* allows pigment)

(c) Use the forked line method for this portion

Combining the phenotypes gives (always count the proportions to see that they add up to 1.0):

> 16/32 albino;
>
> 9/32 gray;
>
> 3/32 yellow;
>
> 3/32 black;
>
> 1/32 cream

(d) Use the forked line method for this portion

Combining the phenotypes gives (always count the proportions to see that they add up to 1.0):

9/16 (gray);

3/16 (black);

4/16 (albino)

(e) Use the forked line method for this portion

```
                      ╱ 1/2 Cc —> 3/8 (gray)
              3/4 B_ ╱╲—1/2 cc —> 3/8 (albino)
    1/1 AA ╱
            ╲  1/4 bb — 1/2 Cc —> 1/8 (yellow)
                       ╲1/2 cc —> 1/8 (albino)
```

The final ratio would be

3/8 (grey);

1/8 (yellow);

4/8 (albino)

18.   Treat each of the crosses as a series of monohybrid crosses, remembering that albino is epistatic to color and black and yellow interact to give cream.

(a) Since this is a 9:3:3:1 ratio with no albino phenotypes, the parents must each have been double heterozygotes and incapable of producing the *cc* genotype.

Genotypes:

AaBbCC   X   AaBbCC

or

AaBbCC   X   AaBbCc

Phenotypes: gray   X   gray.

(b) Since there are no black offspring, there are no combinations in the parents which can produce *aa*. The 4/16 proportion indicates that the *C* locus is heterozygous in both parents. If the parents are

AABbCc   X   AaBbCc

or

AABbCc   X   AABbCc

then the results would follow the pattern given.

Phenotypes: gray   X   gray.

(c) Notice that 16/64 or 1/4 of the offspring are albino, therefore the parents are both heterozygous at the *C* locus. Second, notice that without considering the *C* locus, there is a 27:9:9:3 ratio which reduces to a 9:3:3:1 ratio.   Given this information, the genotypes must be

AaBbCc   X   AaBbCc.

Phenotypes: gray   X   gray

(d) Notice that 2/8 or 1/4 of the offspring are albino which indicates that both parents are *Cc*. Also notice that the ratio of black to cream is 1:1 suggesting that the parents are *Bb* and *bb*. Because there are no grey or yellow offspring, there can be no *A* alleles.

Genotypes:

aaBbCc   X   aabbCc

Phenotypes: black   X   cream.

(e) Notice that half of the offspring are albino indicating that, for the *C* locus the genotypes are *Cc* and *cc*. There is a 3:1 ratio of black to cream indicating heterozygosity for the *B* locus in each parent. Because there are no gray or yellow offspring, there can be no *A* alleles.

Genotypes:

      aaBbCc X aaBbcc

Phenotypes: black X albino

---

19. There are two characteristics being described here; flower color (incomplete dominance since red X white gives all purple and a 1:2:1 ratio in the F_2) and edible portion shape (dominant/recessive).

Symbolism:

    RR = red; *Rr* = purple; *rr* = white

    (refer to K/C-Fig.4.1)

    O = long; *o* = oval

(a) Since true-breeding means homozygous, the genotypes of the parents must be

---

    RROO X rroo.

F_1 = *RrOo* (purple, long)

F_2 = use the forked line method

1/4 red   3/4 long —> 3/16 red, long
         1/4 oval —> 1/16 red, oval

2/4 purple   3/4 long —> 6/16 purple, long
          1/4 oval —> 2/16 purple, oval

1/4 white   3/4 long —> 3/16 white, long
         1/4 oval —> 1/16 white, oval

(b) Notice that there is a 1:1 ratio of red to purple as well as long to oval. This would suggest that the red oval parent was *RRoo* (which is the only genotype possible for this phenotype) and the unknown parent was *RrOo*. This would mean that the phenotype of the unknown parent was purple, long.

20. After reading the problem, glance at the kinds of F_1 and F_2 ratios. Notice that the first two, (A) and (B), appear as monohybrid ratios and (C) is clearly dihybrid. You need to see this combination of crosses as being solved with one set of gene symbols. The fact that cross (c) yields a 9:3:3:1 ratio gives you a start.

(a) Going back to basics, set up the relationship which you know holds for a 9:3:3:1 ratio as follows:

    *A_B_* = 9/16

    A_bb = 3/16

    *aaB_* = 3/16

    aabb = 1/16

Then assign the phenotypes from cross (C) as indicated:

$$A\_B\_ = 9/16 \text{ (green)}$$

$$A\_bb = 3/16 \text{ (brown)}$$

$$aaB\_ = 3/16 \text{ (grey)}$$

$$aabb = 1/16 \text{ (blue)}$$

Now it should become clear that blue results from interaction of the *aa* and *bb* genotypes and brown and grey result from homozygosity of either of the two genes as shown above. From this model, see if crosses (a), (b), and (c) and the resulting progeny make sense.

Cross A:

P$_1$:
            AABB X aaBB

F$_1$:   *AaBB*

F$_2$: *3/4 A_BB: 1/4 aaBB*

Cross B:

P$_1$:
            AABB X AAbb

F$_1$:   *AABb*

F$_2$: *3/4 AAB_: 1/4 AAbb*

Cross C:

P$_1$:
            aaBB X AAbb

F$_1$:   *AaBb*

F$_2$: *9/16 A_B_: 3/16 A_bb:*

    3/16 aaB_: 1/16 aabb

(b) This question is exactly as that in Cross C. The genotype of the unknown P$_1$ individual would be *AAbb* (brown) while the F$_1$ would be *AaBb* (green).

21. First see in this problem that a 9:7 ratio is involved which implies a dihybrid condition with epistasis. Going back to a basic 9:3:3:1 ratio one can see that if the 3:3:1 groups were lumped together the 9:7 ratio would result. Assign tall to any plant with both *A* and *B*, and any dwarf plant which is homozygous for either or both the recessive alleles. The initial cross must have been

$$AABB \quad X \quad aabb.$$

There are two gene pairs involved.

(a)

$$A\_B\_ = 9/16 \text{ (tall)}$$

$$A\_bb = 3/16 \text{ (dwarf)}$$

$$aaB\_ = 3/16 \text{ (dwarf)}$$

$$aabb = 1/16 \text{ (dwarf)}$$

(b) There are three different classes of dwarf plants. Within each of the 3/16 classes there are two types:

$$A\_bb = 3/16 \text{ (dwarf)}$$

$$= 1/3 \ AAbb \text{ and } 2/3 \ Aabb$$

and

$$aaB\_ = 3/16 \text{ (dwarf)}$$

$$= 1/3 \ aaBB \text{ and } 2/3 \ aaBb$$

Therefore the true breeding dwarf plants would be the following:

$$AAbb, \ aaBB, \text{ and } aabb$$

and they would constitute 3/7 of the dwarf group.

22. Problems of this type often cause difficulties for students. It is important for students to go back to basic patterns of inheritance when getting started. First, see that a 9:3:3:1 ratio is involved as indicated below:

$$A\_B\_ = 9/16$$

$$A\_bb = 3/16$$

$$aaB\_ = 3/16$$

$$aabb = 1/16$$

(a) Assign the phenotypes as given, then see if patterns emerge.

$$A\_B\_ = 9/16 \text{ (yellow)}$$

$$A\_bb = 3/16 \text{ (blue)}$$

$$aaB\_ = 3/16 \text{ (red)}$$

$$aabb = 1/16 \text{ (mauve)}$$

From this information, the genotypes for the various phenotypes and the solution to the problem become clear.

As stated in the problem all colors *may* be true-breeding. See that each type can exist as a full homozygote. If plants with blue flowers (homozygotes) are crossed to red-flowered homozygotes, the F_1 plants would have yellow flowers. Also as stated in the problem, if yellow-flowered plants are crossed with mauve-flowered plants, the F_1 plants are yellow and the F_2 will occur in a 9:3:3:1 ratio. All of the observations fit the model as proposed.

(b) If one crosses a true-breeding red plant (*aaBB*) with a mauve plant (*aabb*), the F_1 should be red (*aaBb*). The F_2 would be as follows:

> *aaBb X aaBb* —>
>
> 3/4 *aaB_* (red): 1/4 *aabb* (mauve)

23. This cross is "dihybrid" and involves codominance (AB blood group) and complete dominance (Bombay phenotype). There is also gene interaction in which the *hh* genotype is epistatic over the blood group genes. The mating will be, using symbolism in K/C-Table 4.1,

$$I^A I^B Hh \quad X \quad I^A I^B Hh$$

producing offspring as follows:

1/4 $I^A I^A$ ⟨ 3/4 H_	—>	3/16 type A
1/4 $I^A I^A$ ⟨ 1/4 hh	—>	1/16 type O
2/4 $I^A I^B$ ⟨ 3/4 H_	—>	6/16 type AB
2/4 $I^A I^B$ ⟨ 1/4 hh	—>	2/16 type O
1/4 $I^B I^B$ ⟨ 3/4 H_	—>	3/16 type B
1/4 $I^B I^B$ ⟨ 1/4 hh	—>	1/16 type O

Type O will total 4/16.

24. First, make certain that you understand the genetics of all the gene pairs being described in the problem. The ABO system involves multiple alleles, codominance, and dominance. The MN system is codominant while the Rh system as presented here is a dominant/recessive system.

The easiest way to approach these types of problems is to consider those gene pairs which produce a low number of options in the offspring. Notice in cross #1 there are two options in the offspring for the ABO system (types A and O), but only one option for the MN system (type MN) and one option for the Rh system (Rh⁻). By looking at the most restrictive classes (MN and Rh systems in this case) one can see that option (c) is the only one which is both MN and Rh⁻. The remainder of the combinations can be determined using the same logic.

Cross #1   = (c)
Cross #2   = (d)
Cross #3   = (b)
Cross #4   = (e)
Cross #5   = (a)

There are no other combinations of solutions.

25. (a) Because the denominator in the ratios is 64 one would begin to consider that there are three independently assorting gene pairs operating in this problem. Because there are only two characteristics (eye color and croaking) however, one might hypothesize that two gene pairs are involved in the inheritance of one trait while one gene pair is involved in the other.

(b) Notice that there is a 48:16 (or 3:1) ratio of rib-it to knee-deep and a 36:16:12 (or 9:4:3) ratio of blue to green to purple eye color. Because of these relationships one would conclude that croaking is due to one gene pair while eye color is due to two gene pairs. Because there is a (9:4:3) ratio regarding eye color, some gene interaction (epistasis) is indicated.

(c,d)

Symbolism:

Croaking: $R\_$ = rib-it; rr = knee-deep

Eye color:

Since the most frequent phenotype is blue eye, let $A\_B\_$ represent the genotypes. For the purple class, "a 3/16 group" use the A_bb genotypes. The "4/16" class (green) would be the $aaB\_$ and the $aabb$ groups.

(e) The cross involving a blue-eyed, knee-deep frog and a purple-eyed, rib-it frog would have the genotypes:

$$AABBrr \quad X \quad AAbbRR$$

which would produce an F_1 of $AABbRr$ which would be blue-eyed and rib-it. The F_2 will follow a pattern of a 9:3:3:1 ratio because of homozygosity for the A locus and heterozygosity for both the B and R loci.

$AAB\_R\_$ = blue-eyed, rib-it

AAB_rr = blue-eyed, knee-deep

$AAbbR\_$ = purple-eyed, rib-it

$AAbbrr$ = purple-eyed, knee-deep

(f) The different results can arise because of the genetic variety possible in producing the green-eyed frogs. Since there is no dependence on the B locus, the following genotypes can define the green phenotype:

$$aaBB, \ aaBb, \ aabb.$$

(g) In doing these types of problems, take each characteristic individually, then build the complete genotypes. Notice that the ratio of purple-eyed to green-eyed frogs is 3:1, therefore expect the parents to be heterozygous for the A locus. Because the ratio of rib-it to knee-deep is also 3:1 expect both parents to be heterozygous at the R locus.

The B locus would have the bb genotype because both parents are purple-eyed as given in the problem. Both parents would therefore be $AabbRr$.

26. First, look for familiar ratios which will inform you as to the general mode of inheritance. Notice that the last cross (h) gives a 9:4:3 ratio which is typical of epistasis. From this information one can develop a model to account for the results given.

```
Symbolism:

  A-B-   =   black

  A-bb   =   golden
  aabb   =   golden

  aaB-   =   brown
```

The combination of *bb* is epistatic to the *A* locus.

(a) *AAB- X aaBB* (other configurations possible but each must give all offspring with *A* and *B* dominant alleles)

(b) *AaB- X aaBB* (other configurations are possible but no *bb* types can be produced)

(c) *AABb X aaBb*

(d) *AABB X aabb*

(e) *AaBb X Aabb*

(f) *AaBb X aabb*

(g) *aaBb X aaBb*

(h) *AaBb X AaBb*

Those genotypes which will breed true will be as follows:

black = *AABB*

golden = all genotypes which are *bb*

brown = *aaBB*

27. In this problem, each upper-case letter contributes 5cm above 20cm.

(a) Because ther are four upper-case letters, the plant should be 40cm high.

(b) Since the F_1 plant is AaBb (two upper-case letters), the plant should be 30cm high. There would be 2n+1 (n = number of heterozygous gene pairs) classes of F_2 plants with 0, 1,2,3, and 4 upper-case letters. If you refer to K/C-Fig.4.14 notice in the second example the ratio is 1:4:6:4:1. This should be no mystery because a 16-box Punnett square as in K/C-Fig.3.7 shows the source of all the genotypes.

(c) These are plants with one upper-case letter and three upper-case letters respectively. Think of the alleles in the following manner, $A^1A^2B^1B^2$, with each allele capable of contributing an increment (5cm) to the height of the plant.

The possibilities for a 25cm plant are

$A^1a^2b^1b^2, a^1A^2b^1b^2, a^1a^2B^1b^2,$ and $a^1a^2b^1B^2$

and for a 35cm plant,

$a^1A^2B^1B^2, A^1a^2B^1B^2, A^1A^2b^1B^2,$ and $A^1A^2B^1b^2.$

28. There are many ways to deal with this type of problem. One can make the cross then locate among all the genotypes the one (or several) with only two additive alleles. A simpler approach would be to determine what fraction of the gametes from each parent will, when united, produce a plant with only two additive alleles.

The plant *AaBbCC* will produce four different types of gametes, 1/4 of which will be *abC*. The *AABbcc* individual will produce two different types of gametes, 1/2 of which will be *Abc*. These gametes, when combined, will produce a single genotype with two additive alleles, *AabbCc*, at a frequency of 1/4 X 1/2 = 1/8 of the progeny.

29. As you read this questions, notice that the strains are inbred, therefore homozygous, and that approximately 1/250 represent the shortest and tallest groups in the F_2 generation.

(a, b) Referring to K/C-Fig.4.14, see that where four gene pairs act additively, the proportion of one of the extreme phenotypes to the total number of offspring is 1/256 (add the numbers in each phenotypic class). The same may be said for the other extreme type. The extreme types in this problem are the 12cm and 36cm plants. From this observation one would suggest that there are four gene pairs involved.

(c) If there are four gene pairs, there are nine (2n+1) phenotypic categories and eight increments between these categories (see K/C-Fig.4.14). Since there is a difference of 24cm between the extremes, 24cm/8 = 3cm for each increment (each of the additive alleles).

(d) A typical F_1 cross which produces a "typical" F_2 distribution would be where all gene pairs are heterozygous (*AaBbCcDd*), independently assorting, and additive. There are many possible sets of parents which would give an F_1 of this type. The limitation is that each parent has genotypes which give a height of 24cm as stated in the problem. Because the parents are inbred, it is expected that they are fully homozyogus. An example:

AABBccdd   X   aabbCCDD

(e) Since the *aabbccdd* genotype gives a height of 12cm and each upper-case allele adds 3cm to the height, there are many possibilities for an 18 cm plant:

AAbbccdd,

AaBbccdd,

aaBbCcdd, etc.

Any plant with seven upper-case letters will be 33cm tall:

AABBCCDd,

AABBCcDD,

AABbCCDD, for examples.

30. First, since all of the F_1 plants have the same height, all of the parents must be homozygous. Second, since there *are* nine categories listed , there must be at least 4 (2n+1) gene pairs (see K/C-Fig.4.14). If one assumes four gene pairs, and that the 20 inch plants are at the maximum height, the difference between the extremes is 16 inches (20 inches - 4 inches). Since there are nine phenotypic categories, there are eight increments between these categories (see K/C-Fig.4.14). Dividing 16 inches by 8 gives an increment of 2 inches per category. The parental genotypes would be as follows:

4 inches	=	*aabbccdd*
8 inches	=	*AAbbccdd* (and others)
12 inches	=	*aaBBCCdd* (and others)
16 inches	=	*AABBCCdd* (and others)
20 inches	=	*AABBCCDD*

31. Notice that the F_1 has a stripe length of 22cm and that there are seven categories in the F_2 with 22cm being the intermediate and most frequent phenotype. This symmetry indicates that the F_1 cross involves uniform, fully heterozygous organisms. The number of classes (7) indicates that there are three gene pairs involved (from 2n+1).

That the F_2 extreme phenotypes have a wider range than the parents can easily be explained as long as neither parent is fully homozygous as will be shown later. To substantiate the fact that there are three gene pairs involved, notice that 15/1000 is approximately 1/64, the characteristic denominator of trihybrid crosses.

(a,b) The mode of inheritance is clearly polygenic with three gene pairs acting in an additive fashion. The uniformity of the intervals between the phenotypic classes suggests that all genes involved act in an equal manner.

(c) The increment is 2cm as can be seen by the intervals between each to the varieties in the F_2.

(d) Since there are three gene pairs and the F_1 is fully heterozygous (see above) then the genotype of *AaBbCc* seems appropriate and would give the 22cm stripe length. The parents (20cm and 24cm) to produce the F_1 will be of several possibilities, one example is given here:

AAbbcc   X   aaBBCC

Basically a fully homozygous combination (remember these are inbred strains of skunks) which will produce a fully heterozygous F_1, will work.

(e) A skunk with stripes of 20cm will have two upper-case letters in any combination. Several examples are as follows:

AAbbcc,

AaBbcc,

aaBBcc,

AabbCc, etc.

32. **(a)** There is a fairly continuous range of "quantitative" phenotypes in the F_2 and an F_1 which is between the phenotypes of the two parents; thererfore, one can conclude that some phenotypic blending is occurring which is probably the result of several gene pairs acting in an additive fashion. Because the extreme phenotypes (6cm and 30cm) each represent 1/64 of the total, it is likely that there are three gene pairs in this cross.

Remember, trihybrid crosses which show independent assortment of genes have a denominator (4^3) of 64 in ratios. Also, the fact that there are seven categories of phenotypes, which, because of the relationship 2n+1 = 7, would give the number of gene pairs (n) of 3. The genotypes of the parents would be combinations of alleles which would produce a 6cm (*aabbcc*) tail and a 30cm (*AABBCC*) tail while the 18cm offspring would have a genotype of *AaBbCc*.

(b) A mating of an *AaBbCc* (for example) pig with the 6cm *aabbcc* pig would result in the following offspring:

Gametes (18cm tail)	Gamete (6cm tail)	Offspring
ABC		*AaBbCc* (18cm)
ABc		*AaBbcc* (14cm)
AbC		*AabbCc* (14cm)
Abc	*abc*	*Aabbcc* (10cm)
aBC		*aaBbCc* (14cm)
aBc		*aaBbcc* (10cm)
abC		*aabbCc* (10cm)
abc		*aabbcc* (6cm)

In this example, a 1:3:3:1 ratio is the result. However had a different 18cm tailed-pig been selected a different ratio would occur:

AABbcc   X   aabbcc

Gametes (18cm tail)	Gamete (6cm tail)	Offspring
ABc	*abc*	*AaBbcc* (14cm)
Abc		*Aabbcc* (10cm)

5

Sex Determination and Sex Linkage

Vocabulary: Organization and Listing of Terms

Structures and Substances

Chlamydomonas

    isogametes

    zoospores

Neurospora

    mycelium

    hyphae

    conidia

    mating types (*A, a*)

    protoperithecia

    ascogonium

    trichogynes

    ascus

    ascospores

Zea mays

    gametophyte

    sporophyte

    microgametophyte

    stamen (tassels)

    pistil

    endosperm nucleus

    egg nucleus

    antipodal nuclei

    synergids

    stigma

    diploid zygote

    triploid endosperm

    kernel

Humans

    cortex

    medulla

Sex chromosomes

    X-body

    heterochromosome

    Y-chromosome

        heterochromatin

        euchromatin

    H-Y antigen

histocompatibility antigen

testis-determining factor (TDF)

autosomes

Processes/Methods

Sexual differentiation

    Chlamydomonas

        isogamous

        mating types

    Neurospora

    Zea mays

        double fertilization

        mutation analysis

            tassel seed (*ts*)

            silkless (*sk*)

            barren stalk (*ba*)

Humans

    parallel development

Primary sexual differentiation

Secondary sexual differentiation

Classification

    Unisexual, dioecious, gonochoric

    Bisexual, monoecious, hermaphroditic

    Intersex

    Asexual

Sex determination

    Protenor (XX/XO)

    Lygaeus turicus (XX/XY)

    Melandrium (influence of Y chromosome)

    Drosophila melanogaster

        balance theory: 2A:2A, etc.

        nondisjunction

        aneuploid

            superfemale (metafemale)

            metamale

            intersex

        transformer (*tra*)

    Humans (XY)

        Klinefelters syndrome (47, XXY, *etc.*)

        Turners syndrome (45, XO)

mosaics

    (45, X/46, XY) (45, X/46, XX)

    (47, XXX) syndrome (48, XXXX)

    (49, XXXXX)

    (47, XYY)

    (46,XXp-) (46, XXq-)

    "sex-reversed"

    1A2, ZFY

(XX/XY, ZZ/ZW)

Sex-linkage (X-linked inheritance)

    white in *Drosophila*

    sex chromosomes

    autosomes

    hemizygous

    crisscross pattern

    mosaics

        bilateral gynandromorphs

Holandric inheritance (Y-linked)

Dosage compensation

    sex chromatin body (Barr body)

        N-1 rule

    Lyon hypothesis

        red-green colorblindness

        anhidrotic ectodermal dysplasia

        methylation

Sex -limited inheritance

    cock-feathering

Sex-influenced inheritance

    pattern baldness

    horn formation in sheep

    certain coat patterns in cattle

Concepts

Mating types

    plus (+)

    minus (-)

Sex determination

Sex "related" inheritance (F5.1)

    sex-linked

    sex-limited

    sex-influenced

Heterogametic sex

Homogametic sex

    Sex ratio

        primary

        secondary

F5.1. Illustration of common modes of inheritance which are related to the sex of the organism. Notice that in sex-linked (X-linked) inheritance the gene in question is on the X chromosome. In holandric inheritance, the gene is on the Y chromosome. In sex-limited and sex-influenced inheritance, the gene is on the autosomes.

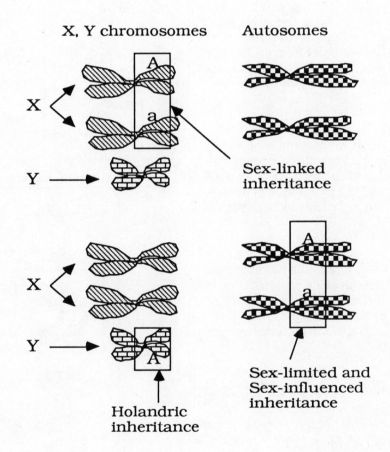

*Solutions to Problems and
Discussion Questions*

1.

(a) *Heterogamy* refers to the condition in which two types of gametes are produced which are related to sex determination. The *heterogametic sex* is that sex in which heterogamy occurs. In humans and *Drosophila*, the male is the heterogametic sex.

(b) The term *heteromorphic chromosome* refers to "different form" of a chromosome. In many cases, the Y chromosome is different from the X chromosome even though they segregate from each other during meiosis. Genes on the heteromorphic chromosome usually play a major role in sex differentiation.

(c) *Nondisjunction* occurs when chromosomes fail to separate normally during mitosis or meiosis. See K/C-Fig.2.12 for an illustration of two types of meiotic nondisjunction. An illustration of mitotic nondisjunction is given in the next column on this page.

(d) A *Barr body* is a darkly staining chromosome seen in some interphase nuclei of mammals with two X chromosomes. There will be one less Barr body than number of X chromosomes. The Barr body is an X chromosome which is considered to be genetically inactive.

(e) The *Lyon Hypothesis* states that the inactivation of the X chromosome occurs at random early in embryonic development. Such X chromosomes are in some way "marked" such that all progeny have the same X chromosome inactivated.

Mitotic nondisjunction is illustrated in the lower panel where the two sister chromatids migrate to the same pole.

Normal mitotic metaphase and anaphase

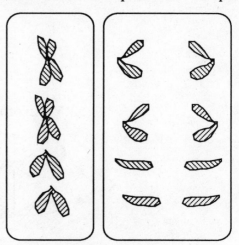

Mitotic nondisjunction

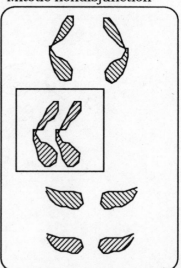

2. Examine the pattern of inheritance of the sex-linked *white-eye* mutation in *Drosophila* in K/C-Fig.5.10 and K/C-Fig.5.11. Notice that there is a chromosomal basis for expression of the recessive *white-eye* gene. This phenomenon is called hemizygosity and it occurs when either there is no homologue for the X chromosome or the existing homologue (the Y chromosome) lacks alleles for genes on the X chromosome. In the following example, the XX/XY symbolsm is used, however one could substitute the ZZ/ZW notation for cross (a). It would also be possible to use the XX/XO notation.

(a)

P$_1$:

    female *rw*/Y X male *RW/RW*

F$_1$:

    1/2 females wild (*RW*/Y),
    1/2 males wild (*RW/rw*)

F$_2$:

    1/4 females reduced wing (*rw*/Y)
    1/4 females wild (*RW*/Y)
    1/2 males wild (*RW/RW* and *RW/rw*)

(b)

P$_1$:

    female *rw/rw* X male *RW*/Y

F$_1$:

    1/2 females wild (*RW/rw*) ,
    1/2 males reduced wing (*rw*/Y)

F$_2$:

    1/4 males reduced wing (*rw*/Y)
    1/4 males wild (*RW*/Y)
    1/4 females wild (*RW/rw*)
    1/4 females reduced wing (*rw/rw*)

(c) With the results given, it would not be possible to distinguish between a *Protenor* and a *Lygaeus* type of sex determination.

3. In the *Lygaeus* type of sex determination, there is an even number of chromosomes in both sexes with one heteromorphic chromosome pair in the male (XY). Presence or absence of the Y chromosome determines sex. In the *Protenor* scheme, there is an even number of chromosomes in the female (XX) but an odd number in the male because the X chromosome has no homologue (XO). Presence or absence of an X chromosome determines sex.

4. Calvin Bridges (1916) studied nondisjunctional *Drosophila* which had a variety of sex chromosome complements. He noted that XO produced sterile males while XXY produced fertile females. In contrast, the Y chromosome was found to be male determining in humans. Individuals with the 47,XXY complement are males while 45,XO produces females.

In *Drosophila* it is the balance between the number of X chromosomes and the number of haploid sets of autosomes which determines sex. In humans there is a small region on the Y chromosome which determines maleness. In *Melandrium* the Y chromosome determines maleness as in humans.

5. The autosomal recessive gene *transformer* (*tra*) causes females to be transformed into sterile males. Normal males are unaffected when they are *tra/tra*.

P$_1$:

    females *sn/sn*; *tra$^+$/tra*

               X

    males *sn$^+$*/Y; *tra$^+$/tra*

F_1:

$sn^+/sn;$ tra^+/tra^+	1/8	females, wild
$sn^+/sn;$ tra^+/tra	2/8	females, wild
$sn^+/sn;$ tra/tra	1/8	sterile **males**, wild
$sn/Y;$ tra^+/tra^+	1/8	males, singed
$sn/Y;$ tra^+/tra	2/8	males, singed
$sn/Y;$ tra/tra	1/8	males, singed

6. Examine K/C-Fig.2.12 and note that in *primary* nondisjunction half of the gametes contain two X chromosomes while the complementary gametes contain no X chromosomes. Fertilization, by a Y-bearing sperm cell, of those female gametes with two X chromosomes would produce the XXY Klinefelters syndrome. Fertilization of the "no-X" female gamete with a normal X-bearing sperm will produce the Turners syndrome.

7. The presence of the Y chromosome provides a factor (or factors) which leads to the initial specification of maleness. Subsequent expression of secondary sex characteristics must be dependent on the interaction of the normal X-linked *Tfm* allele with testosterone. Without such interaction, differentiation takes the female path.

8. There are several possibilities which are discussed in the K/C text. One could account for the significant departures from a 1/1 ratio of males to females by suggesting that at anaphase I of meiosis, the Y chromosome more often goes to the pole which produces the more viable sperm cells. One could also speculate that the Y-bearing sperm has a higher likelihood of surviving in the female reproductive tract, or that the egg surface is more receptive to Y-bearing sperm. At this time the mechanism is unclear.

9. Because synapsis of chromosomes in meiotic tissue is often accompanied by crossing over, it would be detrimental to sex-determining mechanisms to have sex-determining loci on the Y chromosome transferred, through crossing over, to the X chromosome.

10. Many organisms have evolved over millions of years under the fine balance of numerous gene products. Many genes required for normal cellular and organismic function in *both* males and females are located on the X chromosome. These gene products have nothing to do with sex determination or sex differentiation.

At the cellular level, males and females are metabolically similar and apparently the potential differential output in a two X-chromosome cell must be equalized to that of a one X-chromosome cell.

11. There is a simple formula for determining the number of Barr bodies in a given cell: 2N-1, where N is the number of X chromosomes.

Klinefelters syndrome (XXY)	= 1
Turners syndrome (XO)	= 0
47, XYY	= 0
47, XXX	= 2
48, XXXX	= 3

12. Refer to K/C-Figs.5.15-17 and notice that the phenotypic mosaicism is dependent on the heterozygous condition of genes on the two X chromosomes. Dosage compensation and the formation of Barr bodies occurs only when there are two or more X chromosomes. Males normally have only one X chromosome therefore such mosaicism can not occur. Females normally have two X chromosomes. There are cases of male calico cats which are XXY.

13. While there are many theories as to the evolutionary advantage of sexual reproduction, most center on the advantage of organisms having a mechanism for avoiding homozygosity in favor of heterozygosity. In addition, with sexual reproduction, favorable variations (mutations) occurring in one organism may be "shared" with other organisms of the same species. The recombination of genes achieved through sexual mixing occurs at two levels: the *organism* through sexual reproduction and the *chromosome*, through crossing over. Because some species are monoecious there is potential for self-fertilization and therefore the accumulation of homozygosity which is generally associated with fitness depression. See Chapters 24 and 25 in the K/C text. Self-sterility alleles reduce the chances of self-fertilization and enhance the general heterozygosity of the population.

14. In mammals the Y chromosome is male determining. Under its influence testes development is initiated and testicular tissue secretes a hormone that supports male sexual differentiation. Without the Y chromosome, testes development is not initiated, the "male hormone" is not released, and female development is the consequence. When such fraternal twins share circulation, the female embryo comes under the influence of the circulating hormone and masculinization of the female reproductive organization occurs.

15. In a *Drosophila* X chromosome, the centromere is very near one end. Attached-X chromosomes result from an abnormal division of the centromere and sister chromatids remain attached instead of separating at anaphase. The short arm of the X chromosome is lost. Because the centromere is very near the end and the small arm is composed of heterochromatin, attached-X chromosomes contain the genetic equivalent of two X chromosomes. A fly with an attached-X and free Y chromosome (<u>XX</u>Y) is therefore female because the ratio of X chromosomes to the number of haploid sets of autosomes is essentially 1.0. Should a similar event occur in humans, the individual would be male because of the presence of the Y chromosome.

Normal centromere division of chromatids

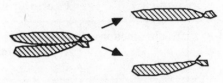

Generation of attached-X chromosomes by abnormal centromere division

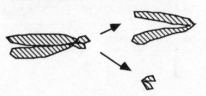

Illustration of the abnormal centromere division which generates attached-X chromosome.

16. Below is an illustration of the behavior of attached-X chromosomes during gamete formation.

Attached-X female

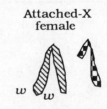

Female gametes produced

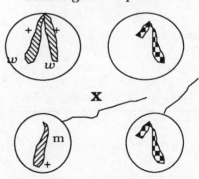

Male gametes

The combinations of the gametes and the phenotypes expected are given below:

Female gametes	Male gametes	Phenotype
Attached-X	X chromosome	dies at third instar larval stage
Attached-X	Y chromosome	white-eyed female
Y chromosome	X chromosome	miniature male
Y chromosome	Y chromosome	dies at egg stage

17. In Turkeys the male is XX, so a nondisjunctional event would produce the possibility of an XX egg which could give a male parthenogenically.

18. Set up the cross in the following manner:

> Xa/Xb female   X   Xa male
>
> 50% fertilized:
>
>   1/4 Xa/Xa   = male
>
>   1/4 Xa/Xb   = female
>
> 50% unfertilized:
>
>   1/4 Xa   = male
>
>   1/4 Xb   = male

19. In mammals, the scheme of sex determination is dependent on the presence of a piece of the Y chromosome. If present a male is produced. In *Bonellia viridis*, the female proboscis produces some substance which triggers a morphological, physiological, and behavioral developmental pattern which produces males.

To elucidate the mechanism, one could attempt to isolate and characterize the active substance by testing different chemical fractions of the proboscis. Secondly, mutant analysis usually provides critical approaches into developmental processes. Depending on characteristics of the organism, one could attempt to isolate mutants which lead to changes in male or female development. Third, by using micro-tissue transplantations, one could attempt to determine which "centers" of the embryo respond to the chemical cues of the female.

20. This problem incorporates X-linked inheritance with mosaicism caused by X chromosome inactivation. The following symbolism is appropriate:

> B = black coat color; b = yellow coat color
>
> Female tortoise-shell = $X^B X^b$ (mosaic)
> Male black   = $X^B Y$

> $X^B X^b$ X $X^B Y$ —>
>
> $X^B X^B$ = female black
>
> $X^B X^b$ = female tortoise-shell
>
> $X^B Y$ = male black
>
> $X^b Y$ = male yellow

Normally there is no way for a tortoise-shell male to be produced, however with nondisjunction of the X chromosome in the female parent, thus producing a gamete containing the two X chromosomes, the coat color genes could be in the heterozygous state and mosaicism may result. The nondisjunctional event must have occurred in meiosis I. If such a female gamete is fertilized by a Y-bearing sperm, then an XXY tortoise-shell male could result.

21. In order to solve this problem one must first see the possible genotypes of the parents and the grandfathers. Since the gene is X-linked the cross will be symbolized with the X chromosomes.

RG = normal vision; rg = color-blind

Mother's father: X^{rg}/Y

Father's father: X^{rg}/Y

Mother: $X^{RG} X^{rg}$

Father: X^{RG}/Y

Notice that the mother must be heterozygous for the rg allele (being normal-visioned and having inherited an X^{rg} from her father) and the father, because he has normal vision, must be X^{RG}. The fact that the father's father is color-blind does not mean that the father will be color-blind. On the contrary, the father will inherit his X chromosome from his mother.

$X^{RG} X^{rg}$ X $X^{RG}/Y \longrightarrow$

$X^{RG} X^{RG}$ = 1/4 daughter normal

$X^{RG} X^{rg}$ = 1/4 daughter normal

X^{RG}/Y = 1/4 son normal

X^{rg}/Y = 1/4 son color-blind

Looking at the distribution of offspring:

(a) 1/4

(b) 1/2

(c) 1/4

(d) zero

22.

The mating is

$$X^{RG}X^{rg}; I^A I^O \quad X \quad X^{RG}Y; I^A I^O$$

Based on the son who is colorblind and blood type O, the mother must have been heterozygous for the RG locus and both parents must have had one copy of the I^O gene. The probability of having a female child is 1/2, that she has normal vision is 1 (because the father's X is normal) and 1/4 type O blood. The final product of the independent probabilities is

$$1/2 \ X \ 1 \ X \ 1/4 \ = \ 1/8$$

23.

Symbolism:

Normal wing margins = sd^+;

scalloped = sd

(a)

P1:

$X^{sd}X^{sd}$ X $X^+/Y \longrightarrow$

F₁:

1/2 X^+X^{sd} (female, normal)

1/2 X^{sd}/Y (male, scalloped)

F₂:

1/4 X^+X^{sd} (female, normal)

1/4 $X^{sd}X^{sd}$ (female, scalloped)

1/4 X^+/Y (male, normal)

1/4 X^{sd}/Y (male, scalloped)

(b)

P1:

   X^+/X^+ X X^{sd}/Y —>

F₁:

1/2 X^+X^{sd} (female, normal)

1/2 X^+/Y (male, normal)

F₂:

1/4 X^+X^+ (female, normal)

1/4 X^+X^{sd} (female, normal)

1/4 X^+/Y (male, normal)

1/4 X^{sd}/Y (male, scalloped)

If the *scalloped* gene were not X-linked, then all of the F₁ offspring would be wild (phenotypically) and a 3:1 ratio of normal to scalloped would occur in the F₂.

24. Assuming that the parents are homozygous, the crosses would be as follows. Notice that the X symbol may remain to remind us that the *sd* gene is on the X chromosome. It is extremely important that one account for both the mutant genes and each of their wild type alleles.

P1:

   $X^{sd}X^{sd}; e^+/e^+$ X $X^+/Y; e/e$ —>

F₁:

1/2 $X^+X^{sd}; e^+/e$ (female, normal)

1/2 $X^{sd}/Y; e^+/e$ (male, scalloped)

F₂:

	X^+e^+	X^+e	$X^{sd}e^+$	$X^{sd}e$
$X^{sd}e^+$				
$X^{sd}e$				
Ye^+				
Ye				

Phenotypes:

3/16   normal females

3/16   normal males

1/16   ebony females

1/16   ebony males

3/16   scalloped females

3/16   scalloped males

1/16   scalloped, ebony females

1/16   scalloped, ebony males

Forked-line method:

P1:

$X^{sd}X^{sd};\ e^+/e^+$ X $X^+/Y;\ e/e$ —>

F_1:

1/2 $X^+X^{sd};\ e^+/e$ (female, normal)

1/2 $X^{sd}/Y;\ e^+/e$ (male, scalloped)

F_2:

	Wings	Color	
1/4	females, normal	3/4 normal	3/16
		1/4 ebony	1/16
1/4	females, scalloped	3/4 normal	3/16
		1/4 ebony	1/16
1/4	males, normal	3/4 normal	3/16
		1/4 ebony	1/16
1/4	males, scalloped	3/4 normal	3/16
		1/4 ebony	1/16

25. It is extremely important that one account for both the mutant genes and each of their wild type alleles.

P1:

$X^+X^+;\ dp/dp$ X $X^w/Y;\ +/+$ —>

F_1:

1/2 $X^+X^w;\ +/dp$ (female, normal)

1/2 $X^+/Y;\ +/dp$ (male, normal)

F_2:

	Eye color	Wing shape	
2/4	females, normal	3/4 normal	6/16
		1/4 dumpy	2/16
1/4	males, normal	3/4 normal	3/16
		1/4 dumpy	1/16
1/4	males, white	3/4 normal	3/16
		1/4 dumpy	1/16

26. Set up the symbolism and the cross in the following manner:

P_1:

$X^+X^+;\ su\text{-}v/su\text{-}v$

X

$X^v/Y;\ su\text{-}v^+/su\text{-}v^+$ —>

F_1:

1/2 $X^+X^v;\ su\text{-}v^+/su\text{-}v$ (female, normal)

1/2 $X^+/Y;\ su\text{-}v^+/su\text{-}v$ (male, normal)

F_2:

2/4	females, $X^+/\_$	3/4 $su\text{-}v^+/\_$
		1/4 $su\text{-}v/su\text{-}v$
1/4	males, X^+/Y	3/4 $su\text{-}v^+/\_$
		1/4 $su\text{-}v/su\text{-}v$
1/4	males, X^v/Y	3/4 $su\text{-}v^+/\_$
		1/4 $su\text{-}v/su\text{-}v$

8/16 wild type females;

（none of the females are homozygouus for the *vermilion* gene)

5/16 wild type males;

(4/4 because they have no *vermilion* gene and 1/4 because the X-linked, hemizygous *vermilion* gene is suppressed by *su-v/su-v*)

3/16 vermilion males;

(no suppression of the *vermilion* gene)

27. It is extremely important that one account for both the mutant genes and each of their wild type alleles.

(a)

P1:

X^vX^v; +/+   X X^+/Y; b^r/b^r —>

F_1:

1/2 X^+X^v; +/b^r (female, normal)

1/2 X^v/Y; +/b^r (male, vermilion)

F_2:

Eye color(X)	Eye color(autosomal)	
1/4 females, normal	— 3/4 normal	3/16
	‾1/4 brown	1/16
1/4 females, vermilion	⌐ 3/4 normal	3/16
	‾1/4 brown	1/16
1/4 males, normal	—3/4 normal	3/16
	‾1/4 brown	1/16
1/4 males, vermilion	—3/4 normal	3/16
	‾1/4 brown	1/16

3/16   = females, normal

1/16   = females, brown eyes

3/16   = females, vermilion eyes

1/16   = females, white eyes

3/16   = males, normal

1/16   = males, brown eyes

3/16   = males, vermilion eyes

1/16   = males, white eyes

(b)

P1:

X^+X^+; b^r/b^r X X^v/Y; +/+   —>

F_1:

1/2 X^+X^v; +/b^r (female, normal)
1/2 X^+/Y; +/b^r (male, normal)

F_2:

Eye color(X)	Eye color(autosomal)	
2/4 females, normal	— 3/4 normal	
	‾ 1/4 brown	
1/4 males, normal	— 3/4 normal	
	‾ 1/4 brown	
1/4 males, vermilion	— 3/4 normal	
	‾ 1/4 brown	

6/16 = females, normal

2/16 = females, brown eyes

3/16 = males, normal

1/16 = males, brown eyes

3/16 = males, vermilion eyes

1/16 = males, white eyes

(c)

P1:

$X^vX^v; \; b^r/b^r \quad X \quad X^+/Y; \; +/+ \; \longrightarrow$

F_1:

1/2 $X^+X^v; \; +/b^r$ (female, normal)
1/2 $X^v/Y; \; +/b^r$ (male, vermilion)

F_2:   Eye color(X)   Eye color(autosomal)

1/4   females,   3/4 normal
      normal   1/4 brown

1/4   females,   3/4 normal
      vermilion   1/4 brown

1/4   males,   3/4 normal
      normal   1/4 brown

1/4   males,   3/4 normal
      vermilion   1/4 brown

3/16 = females, normal

1/16 = females, brown eyes

3/16 = females, vermilion eyes

1/16 = females, white eyes

3/16 = males, normal

1/16 = males, brown eyes

3/16 = males, vermilion eyes

1/16 = males, white eyes

28. In seeing that the distribution of phenotypes in the F_1 is different when comparing males and females, it would be tempting to suggest that the gene is X-linked. However, given that the reciprocal cross gives identical results suggests that the gene is autosomal. Seeing the different distribution between males and females one might consider sex-influenced inheritance as a model and have males more likely to express mahogany and females more likely to express red. This situation is similar to pattern baldness in humans. Consider two alleles which are autosomal and let

---

RR = red, Rr = red in females,

Rr = mahogany in males,

rr = mahogany.

P_1:

female: RR (red)   X   male: rr (mahogany)

F$_1$:

   Rr = females red; males mahogany

   1/2 females (red)

   1/2 males (mahogany)

F$_2$:

   1/4 RR; 2/4 Rr; 1/4 rr

Because half of the offspring are males and half are females, one could, for clarity, rewrite the F$_2$ as:

	1/2 females	*1/2 males*
1/4 *RR*	1/8 red	1/8 red
2/4 *Rr*	2/8 red	2/8 mahogany
1/4 *rr*	1/8 mahogany	1/8 mahogany

29. In looking at the information provided in the text, notice that the only genotype which gives cock-feathering in males is *hh*, while three genotypes give hen-feathering in females,

   HH, Hh, and *hh.*

Remember that these genes are sex-limited and autosomal.

P$_1$:

   female: *HH* X male: *hh*

F$_1$:
      all hen-feathering

F$_2$:

	1/2 females	1/2 males
1/4 HH	hen-feathering	hen-feathering
2/4 Hh	hen-feathering	hen-feathering
1/4 hh	hen-feathering	cock-feathering

All of the offspring would be hen-feathered except for 1/8 males which are cock-feathered.

30. As temperature varies in alligators, and most turtles, various developmental responses occur with respect to sex determination. However, in most lizards, temperature is not influential.

31. The fact that the trait is seen in both males and females rules out the holandric (Y-linked) mode of inheritance. That all of the daughters and none of the sons in generation IV display the trait, suggests that the gene is dominant and on the X chromosome. Daughters inherit one of their X chromosomes from their father while sons inherit their only X chromosome from their mother. Because approximately half of the offspring (including both males and females) in generation V show the trait it is likely that the gene is a dominant.

Linkage, Crossing Over, and Chromosome Mapping

Vocabulary: Organization and Listing of Terms

Structures and Substances

Heterokaryon

Synkaryon

Tetrad

Ascospores

Ascus (pl. asci)

Synaptonemal complex

Endonucleases

Ligases

Bromodeoxyuridine (BUdR)

Processes/Methods

Linkage

   crossing over

      crossover gametes (recombinant)

      recombination

   reciprocal classes

      two-strand stage

      four-strand stage

 complete

   parental (noncrossover gametes)

 incomplete

 chiasmata (chiasma)

 three-point mapping

   product rule (multiple crossovers)

      non-crossovers (NCO)

      single crossovers (SCO)

      double crossovers (DCO)

Determining gene sequence

 correct heterozygous arrangement

 correct sequence of genes

   method I

   method II

distance between gene pairs

discrepancy proportional to distance

Gene to centromere mapping

ordered tetrad analysis

first division segregation

second division segregation

$$\frac{1/2 \text{(second division segregation)}}{\text{total asci scored}}$$

gene conversion

unordered tetrad analysis

parental ditype (P)

nonparental ditype (NP)

tetratypes (T)

$$\frac{NP + 1/2 (T)}{\text{No. tetrads}}$$

Cytological evidence (crossing over)

cytological markers

Mechanism of crossing over

classical theory

chiasmatype theory

late replicating (0.3%) DNA

synaptonemal complex

endonucleases

ligases

zygotene DNA synthesis

pachytene DNA synthesis

Somatic cell hybridization

random loss of human chromosomes

synteny testing

translocation

Haploid organisms

tetrad analysis

Mitotic recombination

synapsis

genetic exchange

twin spots

parasexual cycle

Sister chromatid exchange

bromodeoxyuridine (BUdR)

Bloom syndrome

Mendel and Linkage

independent assortment

Concepts

Linked genes (linkage groups)

arrangement (F6.1)

Chromosome maps

map unit (% recombination)

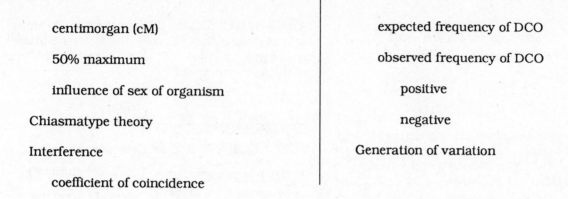

centimorgan (cM)

50% maximum

influence of sex of organism

Chiasmatype theory

Interference

coefficient of coincidence

expected frequency of DCO

observed frequency of DCO

positive

negative

Generation of variation

F6.1. Illustration of two typical configurations (*cis* and *trans*) of two heterozygous gene pairs. Understanding of such arrangements is key to doing linkage problems.

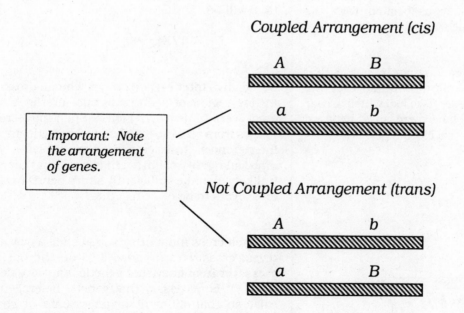

Coupled Arrangement (cis)

A   B

a   b

Important: Note the arrangement of genes.

Not Coupled Arrangement (trans)

A   b

a   B

*Solutions to Problems and
Discussion Questions*

1. The biological significance of genetic exchange and recombination appears to be to generate genetic variation in gametes, thereby leading to genetic variation in organisms. By reshuffling genes, new combinations are generated which may then be of evolutionary advantage. In addition, because chromosomal position can influence gene function, variation is created by *position effect*.

2. First, in order for chromosomes to engage in crossing over, they must be in proximity. It is likely that the side-by-side pairing which occurs during synapsis is the earliest time during the cell cycle that chromosomes achieve that necessary proximity. Second, chiasmata are visible during prophase I of meiosis and it is likely that these structures are intimately associated with the genetic event of crossing over.

3. With some qualification, one can say that crossing over is randomly distributed over the length of the chromosome. Two loci which are far apart are more likely to have a crossover between then than two loci that are close together.

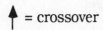

= crossover

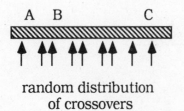

random distribution
of crossovers

4. Because crossing over occurs at the four-strand stage of the cell cycle (that is, after S phase) notice that each single crossover involves only two of the four chromatids.

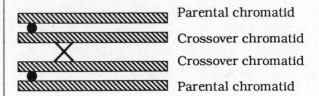

Parental chromatid
Crossover chromatid
Crossover chromatid
Parental chromatid

5. As mentioned in an earlier answer (#3) with some qualifications, crossovers occur randomly along the lengths of chromosomes. Within any region, the occurrence of two events is less likely than the occurrence of one event. If the probability of one event is

$$1/X,$$

the probability of two events occurring at the same time will be

$$1/X^2.$$

6. Positive interference occurs when a crossover in one region of a chromosome interferes with crossovers in nearby regions. Such interference ranges from zero (no interference) to 1.0 (complete interference). Interference is often explained by a physical rigidity of chromatids such that they are unlikely to make sufficiently sharp bends to allow crossovers to be close together.

7. Each cross must be set-up in such a way as to reveal crossovers because it is on the basis of crossover frequency that genetic maps are developed. It is necessary that genetic heterogeneity exist so that different arrangements of genes, generated by crossing over, can be distinguished.

The organism which is heterozygous must be the sex in which crossing over occurs. In other words, it would be useless to map genes in *Drosophila* if the male parent is the heterozygote since crossing over is not typical in *Drosophila* males.

Lastly, the cross must be set-up so that the phenotypes of the offspring readily reveal their genotypes. The best arrangement is one where a fully heterozygous organism is crossed with an organism which is fully recessive for the genes being mapped.

8. Since the distance between *dp* and *ap* is greatest, they must be on the "outside" and *cl* must be in the middle. The genetic map would be as *follows:*

$$dp—cl\text{————————}ap$$
$$\text{3 mu.} \qquad \text{39 mu.}$$

9. The initial cross for this problem would be

$$AaBb \quad X \quad aabb.$$

(a) If the two loci are on different chromosomes, independent assortment would occur and the following distribution (1:1:1:1) is expected:

$$1/4AaBb$$
$$1/4Aabb$$
$$1/4aaBb$$
$$1/4aabb$$

(b) Even though the two loci are linked and on the same chromosome, the frequency of crossing over is so high that crossovers always occur. Under that condition independent assortment would occur and the following distribution (1:1:1:1) is expected:

$$1/4AaBb$$
$$1/4Aabb$$
$$1/4aaBb$$
$$1/4aabb$$

(c) If crossovers never occur, then all of the gametes from the heterozygous parent are *parental.* If the arrangement is

$$AB/ab \quad X \quad ab/ab$$

then the two types of offspring will be

$$1/2AB/ab$$
$$1/2ab/ab.$$

Under this condition *AB* are *coupled.* If, however, *A* and *B* are not coupled then the symbolism would be

$$Ab/aB \quad X \quad aabb.$$

The offspring would occur as follows:

$$1/2Ab/ab$$
$$1/2aB/ab.$$

(d) If the loci are linked with 10 map units between them, then the two recombinant classes, must add up to 10% of the total. Assuming that *A* and *B* are coupled, the following distribution would occur:

45%	*AaBb*	(parental)
5%	*Aabb*	(crossover)
5%	*aaBb*	(crossover)
45%	*aabb*	(parental)

10. In looking at this problem one can immediately conclude that the two loci (kernel color and plant color) are linked because the test cross progeny occur in a ratio other than 1:1:1:1 (and epistasis does not appear because all phenotypes expected are present). The question is whether the arrangement in the parents is *coupled*

$$RY/ry \quad X \quad ry/ry$$

or *not coupled*

$$Ry/rY \quad X \quad ry/ry$$

Notice that the most frequent phenotypes in the offspring, the parentals, are colored, green (88) and colorless, yellow (92). This indicates that the heterozygous parent in the test cross is *coupled*

$$RY/ry \quad X \quad ry/ry$$

with the two dominant genes on one chromosome and the two recessives on the homologue (F6.1). Seeing that there are 20 crossover progeny among the 200, or 20/200, the map distance would be 10 map units (20/200 X 100 to convert to percentages) between the *R* and *Y* loci.

11. Start this problem by working through the expected offspring under two models. One with no crossing over and the second with 30% crossing over in the female.

> **No crossing over:**

Female gametes: Male gametes:

1/2 *e ca⁺*		1/2 *e ca⁺*
1/2 *e⁺ ca*		1/2 *e⁺ ca*

Offspring:

1/4 "e" phenotype

2/4 wild

1/4 "ca" phenotype

> **With 30% crossing over:**

Female gametes: Male gametes:

35% *e ca⁺*		1/2 *e ca⁺*
35% *e⁺ ca*		1/2 *e⁺ ca*
15% *e⁺ ca⁺*		
15% *e ca*		

Offspring:

(obtained by combining gametes and phenotypes)

"e" phenotype   = 17.5% + 7.5%

                  = 25%

wild phenotype   = 17.5% + 7.5% + 17.5% + 7.5%

                  = 50%

"ca" phenotype   = 17.5% + 7.5%

                  = 25%

Notice that the distribution of phenotypes is the same, regardless of the contribution of the crossover classes. Therefore it is impossible to provide the information requested in this problem.

12. Since there is no indication as to the configuration of the P and Z genes (*coupled* or *not coupled*) in the parent, one must look at the percentages in the offspring. Notice that the most frequent classes are PZ and pz. These classes represent the parental (non-crossover) groups which indicates that the original parental arrangement in the test cross was

$$PZ/pz \quad X \quad pz/pz$$

Adding the crossover percentages together

$$(6.9 + 7.1) \text{ gives } 14\%$$

which would be the map distance between the two genes.

13.

	female A:	female B:	Frequency:
NCO	3, 4	7, 8	first
SCO	1, 2	3, 4	second
SCO	7, 8	5, 6	third
DCO	6, 5	1, 2	fourth

The single crossover classes which represent crossovers between the genes which are closer together (d-b) would occur less frequently than the classes of crossovers between more distant genes (b-c).

14. For two reasons, it is clear that the genes are in the *coupled* configuration in the F_1 female. First a completely homozygous female was mated to a wild type male and second, the phenotypes of the offspring indicate the following parental classes

$$sc \ s \ v \text{ and } + + +$$

(a)

P_1:

$$sc \ s \ v \ / sc \ s \ v \quad X \quad + + +/Y$$

F_1:

$$+ + +/sc \ s \ v \quad X \quad sc \ s \ v/Y$$

(b) Using Method I for determining the sequence of genes, examine the parental classes and compare the arrangement with the double crossover (least frequent) classes. Notice that the v gene "switches places" between the two groups (parentals and double crossovers). The gene which switches places is in the middle.

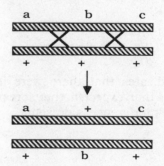

The map distances are determined by first writing the proper arrangement and sequence of genes, then computing the distances between each set of genes.

$$\frac{sc \quad v \quad s}{+ \quad + \quad +}$$

$$sc\text{-}v \quad = \frac{150 + 156 + 10 + 14}{1000} \text{ X } 100$$

$$= 33\% \text{ (map units)}$$

$$v\text{-}s \quad = \frac{46 + 30 + 10 + 14}{1000} \text{ X } 100$$

$$= 10\% \text{ (map units)}$$

Double crossovers are always added into each crossover group because they represent a crossover in each region.

$$sc\text{———}v\text{———}s$$
$$\quad 33 \quad \quad 10$$

(c) The coefficient of coincidence=

$$\frac{\text{observed freq. double C/O}}{\text{expected freq. double C/O}}$$

$$= \frac{(14 + 10)/1000}{.33 \text{ X } .1}$$

$$= \frac{.024}{.033}$$

$$= .727$$

which indicates that there were fewer double crossovers than expected, therefore positive chromosomal interference is present.

15. This set-up involves an F_1 in which the fully heterozygous female has the genes y and w in *coupled* and *ct not coupled*. The arrangement for the cross is therefore:

(a) $y \, w \, +/+ \, + \, ct$ X $y \, w \, +/Y$

It is important at this point to determine the gene sequence. Using Method I, examine the parental classes and compare the arrangement with the double crossover (least frequent) classes. Notice that the w gene "switches places" between the two groups (parentals and double crossovers). The gene which switches places is in the middle. Therefore the arrangement as written above is correct.

(b)

$$y\text{-}w = \frac{9 + 6 + 0 + 0}{1000} \text{ X } 100$$

$$\boxed{= 1.5 \text{ map units}}$$

$$w\text{-}ct = \frac{90 + 95 + 0 + 0}{1000} \text{ X } 100$$

$$\boxed{= 18.5 \text{ map units}}$$

$$y \text{————} w \text{——————————} ct$$
$$0.0 \quad\quad 1.5 \quad\quad\quad\quad\quad 20.0$$

(c) There were

$$.185 \text{ X } .015 \text{ X } 1000 = 2.775$$

double crosovers expected.

(d) Because the cross to the F_1 males included the normal (wild type) gene for *cut wings* it would not be possible to unequivocally determine the genotypes from the F_2 phenotypes for all classes.

16.

(a) The cross will be as follows. Represent the *Dichete* gene as an upper-case letter because it is dominant.

P₁:

D + +/ D + +   X   + e p/+ e p

F₁:

D + +/+ e p X   + e p/+ e p

F₂:

    D + +/+ e p Dichete

    + e p /+ e p ebony, pink

    D e +/+ e p Dichete, ebony

    + + p/+ e p pink

    D + p/+ e p Dichete, pink

    + e +/+ e p ebony

    D e p/+ e p Dichete, ebony, pink

    + + +/+ e p wild type

(b) Using either Method I or Method II determine which gene is in the middle by comparing the parental classes with the double crossover classes. Notice that the *pink* gene "switches places" between the two groups (parentals and double crossovers). The gene which switches places is in the middle. So rewriting the sequence of genes with the correct arrangement gives the following:

F₁:

    D + +/+ p e X   + p e/+ p e

Distances: remember to add in the double crossover classes

$$D\text{-}p = \frac{12 + 13 + 2 + 3}{1000} \times 100$$

$$= 3.0 \text{ map units}$$

$$p\text{-}e = \frac{84 + 96 + 2 + 3}{1000} \times 100$$

$$= 18.5 \text{ map units}$$

17. The fact that two of the genes are linked and 20 map units apart on the third chromosome, and one is on the second chromosome, the problem is a combination of linkage and independent assortment. First provide the genotypes of the parents in the original cross and the reciprocal. Use a semicolon to indicate that two different chromosome pairs are involved.

P₁:

    females: +/+; p e/p e

             X

    males: dp/dp; + +/+ +

F₁:

    females: +/dp; + +/p e

             X

    males: dp/dp; p e/p e

Female gametes: use a modification of the forked-line method for determining the types of gametes to be produced. The *dumpy* locus will give .5 + and .5 *dp* to the gametes because of independent assortment (on a different chromosome) and the other two loci will segregate with 20% (map units) being the recombinants, and 80% being the parentals.

0.5 + — 0.4 + + (parental) = 0.20 + + +
      0.1 + *e* (crossover) = 0.05 + + *e*
      0.1 *p* + (crossover) = 0.05 + *p* +
      0.4 *p* *e* (parental) = 0.20 + *p e*

0.5 *dp* — 0.4 + + (parental) = 0.20 *dp* + +
      0.1 + *e* (crossover) = 0.05 *dp* + *e*
      0.1 *p* + (crossover) = 0.05 *dp p* +
      0.4 *p* *e* (parental) = 0.20 *dp p e*

Crossed with *dp p e* from the male gives the following offspring:

0.20 wild type

0.05 ebony

0.05 pink

0.20 pink, ebony

0.20 dumpy

0.05 dumpy, ebony

0.05 dumpy, pink

0.20 dumpy, pink, ebony

For the reciprocal cross:

F_1:

males: +/*dp*; + +/*p e*

X

females: *dp*/*dp*; *p e*/*p e*

there would be no crossover classes

0.5 +   0.5 + + (parental) = 0.25 + + +
           0.5 *p* *e* (parental) = 0.25 + *p e*

0.5 *dp*   0.5 + + (parental) = 0.25 *dp* + +
           0.5 *p* *e* (parental) = 0.25 *dp p e*

Crossed with *dp p e* from the female gives the following offspring:

.25 wild type

.25 pink, ebony

.25 dumpy

.25 dumpy, pink, ebony

The results would change as a result of no crossing over in males.

18. Since *Stubble* is a dominant mutation (and homozygous lethal) one can determine whether it is heterozygous (*Sb/+*) or homozygous wild type (*+/+*).   One would use the typical test cross arrangement with the *curled* gene so the arrangement would be

$$+ \ cu/ + \ cu$$

19. In typical trihybrid crosses one expects eight kinds of offspring. In this example, only six are listed and one can assume that since the double crossover class is the least frequent, it is the double crossovers which are not listed.

To work this type of problem, examine the list to see which types are not present. In this case, the double crossover classes are the following:

$$+ \ + \ c \ \text{ and } \ a \ b \ +$$

(a,b) Notice that if you compare the parental classes (most frequent) with the double crossover classes (zero in this case) one can, by using the logic of Method I described in the text, determine that the gene *b* is in the middle and the arrangement is as follows. Note: for consistency the zeros are included in the calculations.

$$+ \ b \ c/ a \ + \ +$$

$$a - b \ = \frac{32 \ + \ 38 \ + \ 0 \ + \ 0}{1000} \ \text{X} \ 100$$

$$= 7 \text{ map units}$$

$$b - c \ = \frac{11 \ + \ 9 \ + \ 0 \ + \ 0}{1000} \ \text{X} \ 100$$

$$= 2 \text{ map units}$$

(c) The progeny phenotypes that are missing are *+ + c* and *a b +*, which, of 1000 offspring, 1.4 (.07 X .02 X 1000) would be expected. Perhaps by chance or some other unknown selective factor, they were not observed.

20.

(a) There are several ways to think through this problem. Remember that there is no crossing over in *Drosophila* males therefore any gene on the same chromosome will be completely linked to any other gene on the same chromosome.   Since you can get *pink* by itself, *short* can not be completely linked to it. This leaves linkage to *black* on the second chromosome, the 4th chromosome, or the X chromosome. Since the distribution of phenotypes in males and females is essentially the same, the gene can not be X-linked. In addition, the F_1 males were wild and if the *short* gene was on the X, the F_1 males would be short.

It is also reasonable to state that the gene can not be on the 4th chromosome because there would be eight phenotypic classes (independent assortment of three genes) instead of the four observed.   Through these insights, one could conclude that the *pink* gene is on chromosome 2 with the *black* gene.

Another way to approach this problem is to make three chromosomal configurations possible in the F_1 male. By producing gametes from this male, the answer becomes obvious.

Case A	Case B	Case C
<u>p b</u> <u>sh</u>	<u>p sh b</u>	<u>b sh</u> <u>p</u>
+ + +	+ + +	+ + +

Develop the gametes from Case C and cross them out to the completely recessive triple mutant. You will get the results in the table.

(b) The parental cross is now the following:

Females:	<u>*b s h*</u>	*p*	X	Males:	<u>*b s h*</u>	<u>*p*</u>
	+ +	+			*b s h*	*p*

The new gametes resulting from crossing over in the female would be *b* + and + *sh.* Since the gene *p* is assorting independently, it is not important in this discussion. Because 15% of the offspring now contain these recombinant chromatids, the map distance between the two genes must be 15.

21. The map distance of a gene to the centromere in *Neurospora* is determined by dividing the percentage of second division asci (tetrads) by two. Patterns other than *BBbb* or *bbBB* are "second division" as discussed in the text and represent a crossover between the gene in question and the centromere (see K/C-Fig.6.17c,d). In the data given, the percentage of second division segregation is 20/100 or 20%. Dividing by 2 (because only two of the four chromatids are involved in any single crossover event) gives 10 map units.

22. Because the two types of tetrads occurred at equal frequency, one could say that the gene loci are not linked. Also, looking at the individual loci, notice that gene *a* segregates just as often with gene *b* as it does with its allele +. Because there are no arrangements characteristic of second division segregation (*a+a+*, or *b++b*, for example) the two genes must be very close to their centromeres.

23. The general formula for determination of map distances is as follows:

$$\text{Map distance} = \frac{\text{NP} + 1/2(\text{T})}{\text{Total \# Tetrads}}$$

NP = Nonparental ditypes
T = Tetratypes

For Cross 1:

$$\frac{36 + 14}{100} = 50 \text{ map units}$$

Because there are 50 map units between genes *a* and *b*, they are not linked.

For Cross 2:

$$\frac{3 + 9}{100} = 12 \text{ map units}$$

Because genes *a* and *b* are not linked they could be on non-homologous chromosomes or far apart (50 map units or more) on the same chromosome. Because genes *c* and *b* are linked and therefore on the same chromosome, it is also possible that genes *a* and *c* are on different chromosome pairs. Under that condition, the NP and P (parental ditypes) would be equal; however, there is a possibility that the following arrangement occurs and that genes *a* and *c* are linked.

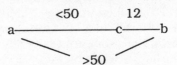

24. By examining K/C-Figs.6.16 and 6.17 you will notice that the only way to get a *BbBb* or the *aa++aa++* ascospore arrangement is if crossing over occurs at the four-strand stage.

25.

(a)

Tetrad	Category
1	parental ditype
2	parental ditype
3	nonparental ditype
4	tetratype
5	tetratype
6	tetratype

(b) If a single crossover occurs (as indicated in the drawing below) between the centromere and the two linked genes, then the arrangement in tetrad 2 will occur.

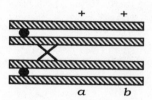

(c)

Map distance $= \dfrac{NP + 1/2(T)}{\text{Total \# Tetrads}}$

NP = Nonparental ditypes
T   = Tetratypes

$= \dfrac{6 + 11}{71} = $ about 24 map units

26. You will notice from K/C.Fig.6.22 that it would take two crossovers to "isolate" the *singed* gene for a spot, and one exchange for a twin spot. Therefore the likelihood of a twin spot is greater than the likelihood of a singed spot. The arrangement of the "markers" being discussed in the second part of this question is as follows:

```
————————t—————————f———c———
        /             |     \
      27.5          56.7    66
```

Because the distance between *forked* and *tan* loci is relatively great, crossovers in this region would be most frequent (and produce a tan spot). The region between the centromere and *forked* is approximately 9 map units, therefore one would expect crossovers to occur in this region at the next highest frequency thereby yielding twin spots. The least frequent event would probably be forked spots because it would take two crossovers to "isolate" the *forked* gene.

27. Because sister chromatids are genetically identical (with the exception of rare new mutations) crossing over between sisters provides no increase in genetic variability. Individual genetic variability could be generated by somatic crossing over because certain patches on the individual would be genetically different from other regions. This variability would be of only minor consequence in all likelihood however. Somatic crossing over would have no influence on the offspring produced .

28. These observations as well as the results of other experiments indicate that the synaptonemal complex is required for crossing over.

29. First make a drawing with the genes placed on the homologous chromosomes as follows:

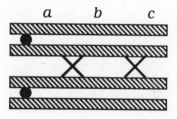

Realize that there are four chromatids in each tetrad and a single crossover involves only two of the four chromatids. The non-involved chromatids must be added to the non-crossover classes. Do all the crossover classes first, then add up the non-crossover chromatids.

For example, in the first crossover class (20 between *a* and *b*) notice that there will be 40 chromatids which were not involved in the crossover. These 40 must be added to the *abc* and +++ classes.

$$
\begin{array}{lll}
a\ b\ c & = & 168 \\
+\ +\ + & = & 168 \\
a\ +\ + & = & 20 \\
+\ b\ c & = & 20 \\
+\ +\ c & = & 10 \\
a\ b\ + & = & 10 \\
+\ b\ + & = & 2 \\
a\ +\ c & = & 2
\end{array}
$$

The map distances woud be computed as follows:

$$
a - b = \frac{20 + 20 + 2 + 2}{400} \ \ \text{X}\ \ 100
$$

$$
= 11 \text{ map units}
$$

$$
b - c = \frac{10 + 10 + 2 + 2}{400} \ \ \text{X}\ \ 100
$$

$$
= 6 \text{ map units}
$$

30.

(a) From K/C-Figs.6.18 and 6.19 one can see the various arrangements and crossover events which are labeled as parental ditype (P), nonparental ditype (NP), and tetratype (T). From that information, the following can be listed:

Tetrad in Problem	Class
1	NP
2	T
3	P
4	NP
5	T
6	P
7	T

(b) The easiest way to determine whether the c and d genes are linked is to compare the frequencies of the parental ditype (P) and nonparental ditype (NP) classes. If P>NP, then the genes are linked. If P=NP they are independently assorting. In the problem given here, P = 44 and NP = 2. Therefore the genes are linked.

(c) The gene to centromere distances are computed by dividing the percent second division segregation by two.

For the centromere to c distance:

$$
\frac{1 + 5 + 3 + 2}{70} = 15.7
$$

now divide by 2 = 7.9 map units

For the centromere to d distance:

$$
\frac{17 + 1 + 3 + 2}{70} = 32.9
$$

now divide by 2 = 16.4 map units

The map would be, according to these figures:

```
C————c————————d
< 7.9 >< 16.4    >
```

or

```
c————————C————————————d
< 7.9 >  <     16.4    >
```

By merely mapping the genes to the centromeres, one can come up with two configurations. Either the genes are on the same side of the centromere or they are on opposite sides. To decide which configuration is occurring, one should see what type of crossovers are required in each case. The two tetrad arrangements which are critical in dealing with the above configurations are #4 and #5.

Notice that, for tetrad arrangement #4 it takes three crossovers if genes *c* and *d* are on the same side of the centromere while it takes two crossovers if they are on opposite sides. For tetrad arrangement #5 it takes two crossovers if genes *c* and *d* are on the same side of the centromere while it takes three crossovers if they are on opposite sides.

Notice that the frequency of tetrad arrangement #5 (5) is much greater than tetrad arrangement #4 (1). This would make sense if the genes were on the same side of the centromere because the highest number of tetrads (5) is in tetrad arrangement #5 where the lowest number of crossovers is required.

If the two genes were on opposite sides of the centromere, the highest number of tetrad arrangements (5 in #5) would be associated with three crossover events. Since the likelihood of multiple crossovers decreases as the number of crossovers increases, it would seem reasonable that the genes are on the same side of the centromere.

Tetrad class #4

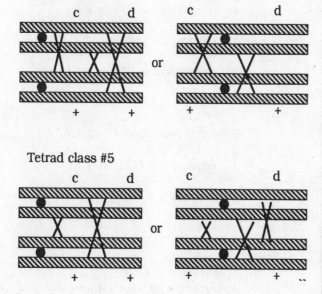

Tetrad class #5

(d) The formula for calculating the distance between the two loci is as follows:

$$\text{Map distance} = \frac{NP + 1/2(T)}{\text{Total \# Tetrads}}$$

NP = Nonparental ditypes
T = Tetratypes

$$= \frac{2 + 12}{70} = 20 \text{ map units}$$

(e)

C————c————d
< 7.9 >
<   16.4   >

The discrepancy between the two mapping systems is caused by the manner in which first and second division segregation products are scored. For instance, in tetrad arrangement #1 there are actually two crossovers between the *d* gene and the centromere, but it is still scored as a first division segregation. In tetrad arrangement #4, three crossovers occur between the *d* gene and the centromere but they are scored as one. If one draws out all the crossovers needed to produce the tetrad arrangements in this problem it would become clear that there are many crossovers between the *d* gene and the centromere which go undetected in the scoring of the arrangements of the *d* gene itself. This will cause one to underestimate the distance and give the discrepancy noted. One could account for these additional crossover classes to make the map more accurate.

(f)

Tetrad class #6

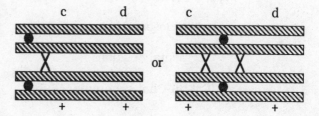

7

Variation in Chromosome Number and Arrangement

Vocabulary: Organization and Listing of Terms

Structures and Substances

Colchicine

Telomeres

rDNA

Nucleolar organizer (NOR)

Micronuclei

Processes/ Methods

Chromosome mutations (aberrations)

    aneuploidy(F7.1)

        XX/XO

        Turners syndrome (humans)

        haplo-VI (*Drosophila*)

        partial monosomy

            (segmental deletions)

            (Cri-du-Chat syndrome, 46,5p-)

trisomy

    XXX (*Drosophila*, humans)

    Datura

    pairing configurations

        trivalent

    Down syndrome (G group)

        trisomy 21 (47, 21+)

    Patau syndrome (D group)

        trisomy 13 (47, 13+)

    Edwards syndrome (E group)

        trisomy 18 (47, 18+)

reduced viability

    gametes

    embryos

    spontaneously aborted fetuses

euploidy(F7.1)

monoploid (n)

diploid (2n)

polyploid

    triploid (3n)

    tetraploid (4n)

    pentaploid (5n)

autopolyploidy

    autotriploids (3n)

        complete nondisjunction

        dispermic fertilization

        tetraploid X diploid

    autotetraploids

      cold or heat shock

      colchicine

allopolyploidy

    hybridization

    allotetraploid (amphidiploid)

      (cotton, *Triticale*)

    protoplast fusion

endopolyploidy

    endomitosis

nondisjunction

    maternal, paternal

    amniocentesis

    chorionic villus sampling (CVS)

translocation

    familial Down syndrome

Chromosome structure

    deletions

      terminal, intercalary

      loop (deficiency, compensation)

      pseudodominance

    duplications

      gene redundancy

        rDNA

        bobbed

      gene amplification

        nucleolar organizer (NOR)

        micronucleoli

      position effect

    rearrangements

      inversions

        paracentric

        pericentric

        arm ratio

        heterozygotes

        unusual pairing arrangments

          inversion loops

          dicentric chromatids

          acrocentric chromatids

          dicentric bridges

"suppression of crossing over"

position effect

translocations

   reciprocal

   unorthodox synapsis

   semisterility

familial Down syndrome

   centric fusion

      Robertsonian fusion

      14/21 or D/G

fragile sites

   X chromosome

   Martin-Bell syndrome (MBS)

Concepts

Significance
(variation in chromosome number)

   genomic balance

      plant vs. animal survival

   sex chromosome balance

   rapid evolution

Gene duplication (evolutionary aspects)

   sequence homology

      nucleic acids

      amino acids

Inversions

   "suppression of crossing over"

   evolutionary consequences

F7.1. Illustration of the chromosomal configurations of duploid and aneuploid genomes of *Drosophila melanogaster.*

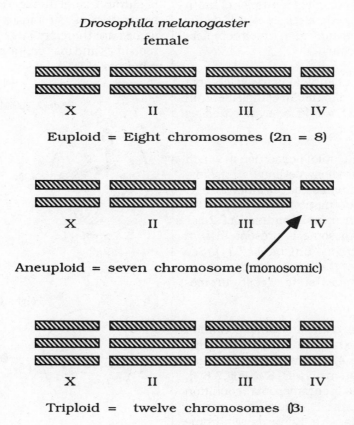

Solutions to Problems and Discussion Questions

1. With frequent exceptions especially in plants, organisms typically inherit one chromosome complement (*haploid* = *n* = one representative from each homologous pair of chromosomes) from each parent. Such organisms are *diploid*, or 2n. When an organism contains complete multiples of the *n* complement (3n, 4n, 5n, etc.) it is said to be *euploid* in contrast to *aneuploid* in which complete haploid sets do not occur.

An example of an aneuploid is *trisomic* where a chromosome is added to the 2n complement. In humans, a trisomy 21 would be symbolized as 2n+1 or 47,21+.

Monosomy is an aneuploid condition in which one member of a chromosome pair is missing, thus producing the chromosomal formula of 2n-1. Haplo-IV is an example of monosomy in *Drosophila*. *Trisomy* is the chromosomal condition of 2n+1 where an extra chromosome is present. Down syndrome is an example in humans (47, 21+). See K/C-Fig.7.4 and notice that all the chromosomes are present in the diploid state except chromosome #21.

Patau syndrome is a chromosomal condition where there is an extra D group chromosome. Such individuals are 47,13+ and have multiple congenital malformations (see K/C-Fig.7.6). *Edwards syndrome* is a chromosomal condition where there is an extra E group chromosome (47,18+). Individuals with Edwards syndrome have multiple congenital malformations and reduced life expectancy (see K/C-Fig.7.7).

Polyploidy refers to instances where there are more than two haploid sets of chromosomes in an individual cell. *Autopolyploidy* refers to cases of polyploidy where all the chromosomes in the individual originate from the same species.

Allopolyploidy involves instances where the chromosomes originate from the hybridization of two different species, usually closely related. Carefully examine K/C-Fig.7.9.

Endopolyploidy is a condition produced by chromosomal replication and separation, but without the typical formation of daughter nuclei (cells). Occurring in specialized cells, endopolyploidy results in cells with 4n, 8n, or 16n for examples. Polytene (see K/C-Fig.7.12) is a special situation of chromosomal duplication, however, there is no separation of chromosomes. Polytene chromosomes are therefore larger than metaphase chromosomes and may contain many times the normal amounts of DNA.

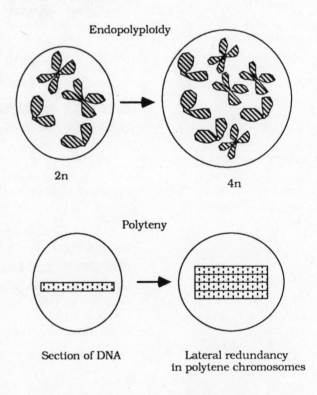

Endopolyploidy

2n   4n

Polyteny

Section of DNA   Lateral redundancy in polytene chromosomes

Notice in K/C-Fig.7.16 the difference between *paracentric* and *pericentric* inversions. In paracentric inversions, the centromere is not included in the region bounded by the breakpoints whereas in pericentric inversions, the breakpoints include the centromere.

2. Individuals with Down syndrome, while suffering congenital defects, tendencies toward respiratory disease, and leukemia, can live well into adulthood. Individuals with Patau or Edwards syndrome live less than four months on the average. Comparing the different sizes of the involved chromosomes (21, 13, and 18, respectively) in K/C-Fig.7.4 for example, suggests that the larger the chromosome, the lower the likelihood of lengthy survival. In addition, it would be expected that certain chromosomes, because of their genetic content, may have different influences on development.

3. The fact that there is a significant maternal age effect associated with Down syndrome indicates that nondisjunction in older females contributes disproportionately to the number of Down syndrome individuals. In addition, certain genetic and cytogenetic marker data indicate the influence of female nondisjunction.

4. While several trisomies (for chromosomes 21, 18, 13, the X and Y) are tolerated, monosomy for the autosomes is not tolerated. The delicate genetic balance produced by millions of years of evolution must be maintained in order for any organism (but especially animals) to develop normally. Monosomy leads to the exposure of recessive, deleterious genes thus producing developmental abnormalities.

Dosage compensation of the sex chromosomes and the relative paucity of Y-linked genes probably contributes to the survival of sex-chromosome aneuploidy. Notice how large the X chromosome is compared with other chromosomes. See K/C-Fig.7.4.

5. As stated in the text, at least 20 percent of all conceptions are terminated in natural abortion. Of these, thirty percent show some chromosomal anomaly. Of the chromosomal anomalies that occur, approximately ninety percent are eliminated by spontaneous abortion. Aneuploidy contributes to the majority of spontaneous abortions.

Trisomy for every human chromosome has been observed, however, monosomy, the reciprocal meiotic event of trisomy, is rare. This observation probably results from gamete or early embryonic inviability.

6. There are numerous examples of autopolyploid plant species which reproduce successfully. As stated in the text, "those with even numbers of haploid genomes are quite successful in sexual reproduction." Allopolyploidy provides a mechanism for relatively rapid evolution because new potentially fertile organisms can be produced in a relatively short period of time. If a "chromosome doubling" step occurs and if there is little homology among chromosomes of the two parental species, then fertility is possible assuming an appropriate developmental environment exists.

7. For largely unknown reasons, plants tolerate increased levels of ploidy more than do animals. Polyploidy provides for greater phenotypic and genetic variation in plants, where concomitant reduction in fertility is overcome by various forms of asexual reproduction. Alloploidy provides an exceptional level of genetic mixing by allowing genomes from different species to interact in the formation of a new species. The paucity of polyploidy in animals may result from unknown developmental consequences mentioned above as well as upsets in mechanisms of sex determination.

8. The sterility of interspecific hybrids is often caused from a high proportion of univalents in meiosis I. As such, viable gametes are rare and the likelihood of two such gametes "meeting" is remote. Even if partial homology of chromosomes allows some pairing, sterility is usually the rule. The horticulturist may attempt to reverse the sterility by treating the sterile hybrid with colchicine. Such a treatment, if sucessful, may double the chromosome number and each chromosome would now have a homologue with which to pair during meiosis.

Normal bivalents
at metaphase I

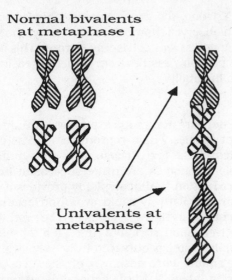

Univalents at
metaphase I

The above diagram refers to the discussion in answer for question 8.

9. Temperature shock applied during meiosis or colchicine applied during mitosis may lead to chromosome doubling. Colchicine interferes with spindle fiber formation. See K/C-Fig.7.8.

10. Because of various types of segregation of chromosomes in triploids (see K/C-Fig.7.3) gametes carrying various combinations of chromosomes (n+1, n+2, etc.) can often be produced. Combinations of such gametes provided the variety for such experiments.

11. American cultivated cotton has 26 pairs of chromosomes; 13 large, 13 small. Old world cotton has 13 pairs of large chromosomes and American wild cotton has 13 pairs of small chromosomes. It is likely that an interspecific hybridization occurred followed by chromosome doubling. These events probably produced a fertile amphidiploid (allotetraploid). Experiments have been conducted to reconstruct the origin of American cultivated cotton.

12. If the diploid chromosome number is 18, 2n=18, then in the somatic nuclei of

> haploid individuals n = 9,
>
> triploid (3n) = 27,
>
> tetraploid (4n) = 36.

A trisomic has one extra chromosome, therefore it is 2n+1 = 19, and a monosomic is 2n-1 = 17. Refer to F7.1.

13. Examine K/C-Fig.7.11 and note the models for the formation of both terminal and intercalary deletions. Such deletions may also be caused by oblique synapsis and crossing over as illustrated in K/C-Figs.7.13 and 7.14(b). Duplications may also be caused by this process as shown. In addition, an error in replication, perhaps by some "rocking motion" of a polymerase may cause duplication.

Inversions may form in the process of loop formation and subsequent double breaks and healing (rejoining). See K/C-Fig.7.15. Perhaps a crossover-like event occurs in the place of overlap. Translocations may originate by the proximity of non-homologous chromosomes and subsequent breakage and rejoining as illustrated in K/C-Fig.7.18.

14. Basically the synaptic configurations produced by chromosomes bearing a deletion or duplication (on one homologue) are very similar. There will be point-for-point pairing in all sections which are capable of pairing. The section which has no homologue will "loop out" as in K/C-Fig.7-12.

15. Inversion "heterozygotes" form loops such that the "outside portion of the loop" is the standard arrangement and the "inside portion of the loop" forms from the inverted homologue. Parts (a) and (b) of K/C-Fig.7.17 illustrate these points.

Translocation "heterozygotes" pair in such a way as to bring homologous segments into synapsis. In part (b) of K/C-Fig.7.18. notice that all homologous sections are in contace through the formation of a "cross." See the answer for question #27.

16. While there is the appearance that crossing over is suppressed in inversion "heterozygotes" the phenomenon extends from the fact that the crossover chromatids end up being abnormal in genetic content. As such they fail to produce viable (or competitive) gametes or lead to zygotic or embryonic death. Notice in K/C-Fig.7.17, the crossover chromatids end up genetically unbalanced.

17. Examine K/C-Fig.7.17 and notice that in (a) there are two genetically balanced chromatids (normal and inverted) and two, those resulting from a single crossover in the inversion loop, which are genetically unbalanced and abnormal (dicentric and acentric). The dicentric chromatid will often break, thereby producing highly abnormal fragments whereas the acentric fragment is often lost in the meiotic process. In part (b) all the chromatids have centromeres, but the two chromatids involved in the crossover are genetically unbalanced. The balanced chromatids are of normal or inverted sequence.

18. Considering that there are at least three map units between each of the loci, and that only four phenotypes are observed, it is likely that genes *a b c d* are included in an inversion and crossovers which do occur among these genes are not recovered because of their genetically unbalanced nature. In a sense, the minimum distance between loci *d* and *e* can be estimated as 10 map units

$$(48 + 52/1000);$$

however this is actually the distance from the *e* locus to the breakpoint which includes the inversion.

The "map" is therefore as drawn below:

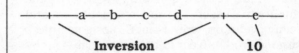

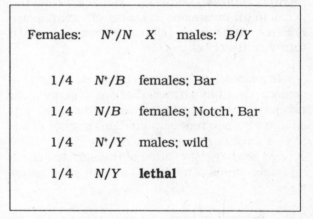

19. The mutant *Notch* in *Drosophila* produces flies with abnormal wings. It is a sex-linked dominant gene (a deletion) which also behaves as a recessive lethal. A deficiency (compensation) loop indicates that bands 3C2 through 3C11 are involved. Loci near *Notch* display pseudodominance. The *Bar* gene on the other hand results from a duplication of a sex-linked region (16A) and results in abnormal eye shape.

Females:	N^+/N X	males: B/Y
1/4	N^+/B	females; Bar
1/4	N/B	females; Notch, Bar
1/4	N^+/Y	males; wild
1/4	N/Y	**lethal**

The final phenotypic ratio would be 1:1:1 for the phenotypes shown above.

20. If a gene exists in a duplicated state and if that gene's product is required for survival, then mutation in either (but not both) the original gene or its duplicate will not ordinarily threaten the survival of the organism. Duplication of a gene provides a buffer to mutation.

21. In a work entitled *Evolution by Gene Duplication*, Ohno, suggests that gene duplication has been essential in the origin of new genes. If gene products serve essential functions, mutation and therefore evolution, would not be possible unless these gene products could be compensated for by products of duplicated, normal genes.

The duplicated genes, or the original genes themselves, would be able to undergo mutational "experimentation" without necessarily threatening the survival of the ogranism.

22. It is likely that when certain combinations of genes are of selective advantage in a specific and stable environment, it would be beneficial to the organism to protect that gene combination from disruption through crossing over. By having the genes in an inversion, crossover chromatids are not recovered and therefore are not passed on to future generations.

Translocations offer an opportunity for new gene combinations by associations of genes from nonhomologous chromosomes. Under certain conditions such new combinations may be of selective advantage and meiotic conditions have evolved so that segregation of translocated chromosomes yields a relatively uniform set of gametes.

23. A Turner syndrome female has the sex chromosome composition of XO. If the father had hemophilia it is likely that the Turner syndrome individual inherited the X chromosome from the father and no sex chromosome from the mother. If nondisjunction occurred in the mother, either during meiosis I or meiosis II, an egg with no X chromosome can be the result.

See K/C-Fig.2.12 for a diagram of primary and secondary nondisjunction.

24. The primrose, *Primula kewensis*, with its 36 chromosomes, is likely to have formed from the hybridization and subsequent chromosome doubling of a cross between the two other species, each with 18 chromosomes. An example of this type of allotetraploidy (amphidiploidy) is seen in K/C-Fig.7.9.

25. Given the basic chromosome set of nine unique chromosomes (a haploid complement) other forms with the "n multiples" are forms of autotetraploidy. In the illustration below the *n* basic set is multiplied to various levels as is the autotetraploid in the example.

Basic set of nine unique chromosomes (n)

Autotetraploid (4n)

Individual organisms with 27 chromosomes (3n) are more likely to be sterile because there are trivalents at meiosis I which caused a relatively high number of unbalanced gametes to be formed.

26. The rare double crossovers in the boundaries of a paracentric or pericentric inversion produce only minor departures from the standard chromosomal arrangement as long as the crossovers involve the same two chromatids. With two-strand double crossovers, the second crossover negates the first. However, three-strand and four-strand double crossovers have consequences which lead to anaphase bridges as well as a high degree of genetically unbalanced gametes.

27. As seen in the diagram below, the homologous sections of the chromosomes can pair by this "cross" configuration. Notice however, that one of the chromosomes is acentric while the other is dicentric. The acentric section will probably not be included in daughter nuclei and will therefore be lost. Depending on the manner in which the spindle fibers attach and segregation occurs, gametes will be variously unbalanced. Notice the segregation patterns in K/C-Fig.7.18(c).

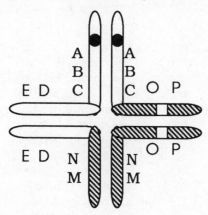

28. Notice in the drawing below that if centromeres #1 and #3 go to the same pole, a balanced gamete will be formed. The other product from this segregation will be a dicentric chromosome which is unbalanced and will not function. All other segregational circumstances lead to unbalanced gametes. Therefore the highest probability for a balanced gamete to be formed is 0.25.

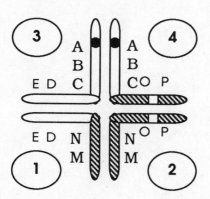

29. Set up the cross in the usual manner, realizing that recessive genes in the Haplo-IV individual will be expressed.

Let b = bent bristles; b^+ = normal bristles

(a)

_/b X b^+/b^+ ——>	

F_1:

_/b^+	= normal bristles
b/b^+	= normal bristles

F_2:

_/b^+ X b/b^+ ——>	

_/b^+	= normal bristles
_/b	= bent bristles
b^+/b^+	= normal bristles
b/b^+	= normal bristles

(b)

_/b^+ X b/b ——>	

F_1:

_/b	= bent bristles
b/b^+	= normal bristles

F_2:

_/b X b/b^+ ——>	

_/b^+	= normal bristles
_/b	= bent bristles
b^+/b	= normal bristles
b/b	= bent bristles

30. In the trisomic, segregation will be "2 X 1" as illustrated below:

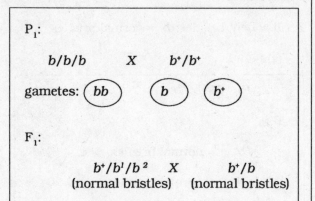

P_1:

$b/b/b$ X b^+/b^+

gametes: (bb)   (b)   (b^+)

F_1:

$b^+/b^1/b^2$ X b^+/b
(normal bristles)   (normal bristles)

Notice that there are several segregation patterns created by the trivalent at anaphase I.

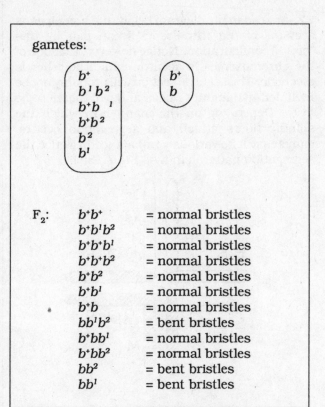

gametes:

b^+
$b^1 b^2$
$b^+ b^1$
$b^+ b^2$
b^2
b^1

b^+
b

F_2:

b^+b^+	= normal bristles
$b^+b^1b^2$	= normal bristles
$b^+b^+b^1$	= normal bristles
$b^+b^+b^2$	= normal bristles
b^+b^2	= normal bristles
b^+b^1	= normal bristles
b^+b	= normal bristles
bb^1b^2	= bent bristles
b^+bb^1	= normal bristles
b^+bb^2	= normal bristles
bb^2	= bent bristles
bb^1	= bent bristles

Part One: Sample Test Questions (with detailed explanations of answers)

> The purpose of these Sample Test Questions is to present a slightly different style of question. Set aside approximately four hours of study time, perhaps a week before the first examination. Attempt to work these questions, five at a time, under test conditions. Write down your answers on paper, then, *after* you have finished the "test," check your answers. If you are having difficulty, then you are weak in the concept areas listed for each question. If you have made mistakes, take comfort, there are many places to make mistakes on these problems. Some of the students who made the same mistakes are now practicing geneticists! Assuming that the first examination covers Chapters 1 through 7, these questions should apply.

Question 1. The mosquito, *Culex pipiens*, has a diploid chromosome number of 6. Assume that one chromosome pair is metacentric, and the other two pairs are acrocentric.

(a) Draw chromosomal configurations which one would expect to see at the following stages: primary oocyte (metaphase I), secondary spermatocyte (metaphase II).

(b) Assuming that a G_1 nucleus in *Culex* contains about 20 picograms (pg) of DNA, how much DNA would you expect in the following nuclei; Primary Spermatocyte, First Polar Body, Secondary Oocyte, Ootid in G_1 phase.

(c) Assume that a female mosquito is heterozygous for the recessive gene *wavy bristles* (symbolized as *wb*) and this gene locus is on an acrocentric chromosome. Draw an expected mitotic metaphase with the appropriate genetic labeling pattern.

> **Concepts:**
>
> **chromosome mechanics**
>
> **mitosis, meiosis**
>
> **symbolism**
>
> **DNA content (cell cycles)**

Answer 1. This question is intended to determine your understanding of mitosis, chromosome morphology, symbolism, the positioning of genes on chromosomes, and the changes in DNA content through the cell cycles.

(a) Since the diploid chromosome number is six, there will be three bivalents, one involving metacentrics, and two involving acrocentrics in a primary oocyte. We can draw the chromosomes of the primary oocyte as follows:

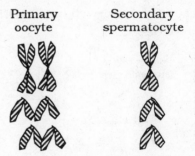

Primary oocyte   Secondary spermatocyte

Note: homologous chromosomes will have "slashes" going in opposite directions because of the shadowing used to depict maternal and paternal chromosomes.

Since secondary spermatocytes arise after meiosis II, there should be only dyads, one metacentric and two acrocentric, and they should be aligned end-to-end as indicated in the drawing above.

(b) Given that there are about 20 picograms of DNA in a G_1 nucleus, we would expect there to be 40pg in a G_2 nucleus (after S phase) and 40pg to the point where homologous chromosomes separate in meiosis I. Secondary spermatocytes and secondary oocytes (as well as first polar bodies) should therefore, each have 20pg of DNA. After meiosis II the resulting nuclei should have 10pg each. If you understand events at interphase and in meiosis, this question is easy to answer. Carefully examine the figure below to understand events during the interphase as far as DNA content is concerned. Then examine F2.3 to see how chromosomes are behaving in meiosis. From this information you should see the answers as follows:

Primary spermatocyte = 40 pg

First Polar Body   = 20pg

Secondary oocyte   = 20pg

Ootid (in G_1)   = 10pg

It might be helpful to view changes in DNA content in graphic form:

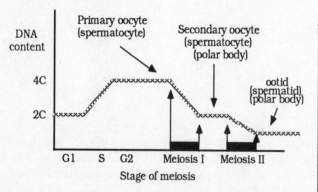

The "C" stands for "complements" of DNA.

(c) If the mosquito is heterozygous for the recessive gene *wavy bristles* (*wb*) then it would have the genotype *Wb/wb*. Because there are four letters here representing the two genes, the slash between the symbols helps us to understand that there are only two genes being discussed.

We are asked to draw an acrocentric, mitotic metaphase chromosome complement in this heterozygous insect. We are expected to place the gene symbols on the chromosomes. Recall that there is no synapsis of homologous chromosomes in mitotic cells, therefore, the chromosomes should not be placed side-by-side. Since sister chromatids are *identical* and homologous chromosomes are *similar*, we should draw the figure as follows:

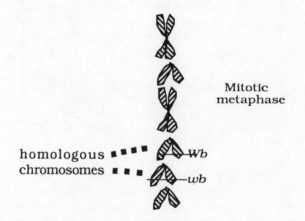

Mitotic metaphase

homologous chromosomes

Wb

wb

Common errors:

   incorrect number of chromosomes

   incorrect chromosome morphology

      metacentric, acrocentric

   **poor relationship of DNA
      content to cells**

   inappropriate symbols

   inappropriate placement of genes

Question 2. In humans, chromosome #1 is large and metacentric, the X chromosome is medium in size and submetacentric (submedian), while the Y chromosome is small and acrocentric. Assume that you were microscopically examining human chromosomes at the stages given below.

(a) Illustrate (draw) the above-mentioned (#1, X and/orY) chromosomes and/or pairs at the stages given (several different configurations may be applicable in some cases):

Metaphase I (Primary Oocyte):

First Polar Body:

Secondary Spermatocyte:

Secondary Oocyte:

(b) The Rh blood group locus is on Chromosome #1. Individuals are *DD* or *Dd* if Rh$^+$ and *dd* if Rh$^-$. The locus for glucose-6-phosphate-dehydrogenase deficiency (G6PD) is located on the X chromosome. There are two alternatives at this locus, + and -. For each of the above cells, place genes (using symbolism given) on chromosomes if the female is heterozygous at both the Rh and G6PD loci. Do the same for the secondary spermatocyte (above) assuming that the male is Rh-, and + for the G6PD locus.

(c) Assume that the average DNA content per G_1 nucleus in humans is 6.5 picograms. For the nuclei (including the entire chromosome complement for each nucleus) presented, give the expected DNA content:

Metaphase I (Primary Oocyte) :

First Polar Body:

Secondary Spermatocyte:

Secondary Oocyte:

Concepts:

chromosome mechanics

mitosis, meiosis

symbolism

DNA content (cell cycles)

Answer 2. (a,b) Recall that a metacentric chromosome has "arms" of approximately equal length, while submetacentric and acrocentric chromosomes have arms of unequal length (see K/C-Fig. 2.3).

Metaphase I (Primary Oocyte): homologous chromosomes are replicated and synapsed. There will be two X chromosomes present because oocytes occur in females. On the metacentric chromosomes (#1), place, such that sister chromatids are identical, the *Dd*. Place the + - alternatives on the X chromosomes.

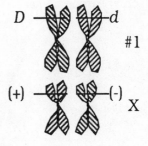

First Polar Body: the first polar body is a product of meiosis I, after homologous chromosomes have migrated to opposite poles. At this stage, dyads are present. Because females produce polar bodies, there should be an X chromosome present. Because the female is heterozygous, there are several possible answers. Note that there is only one representative of each allele for each gene pair.

One possibility

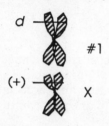

Secondary Spermatocyte: a secondary spermatocyte will have the same chromosome configuration as a First Polar Body. Both are products of meiosis I and dyads should be present. Because spermatocytes occur in males, there will either be an X chromosome or a Y chromosome present. There are therefore two possible answers. Regarding the genetic constitution of these cells, as stated in the problem, we are to assume that the male is Rh⁻ and + for the G6PD locus. Since this locus is on the X chromosome, only one genotype (regarding the X chromosome) can be presented.

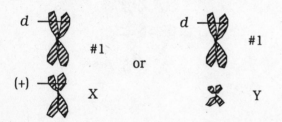

Secondary Oocyte: being a product of meiosis I, dyads will be present. Because oocytes occur in females, there should be an X (not a Y) chromosome present. The genetic labeling pattern for the secondary oocyte will be the same as for the first polar body.

One possibility

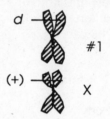

(c) Referring to the figure for part (b) in the above problem, if a G_1 nucleus contains 6.5pg DNA, then the following DNA contents are expected.

Metaphase I (Primary Oocyte):	13pg
First Polar Body:	6.5pg
Secondary Spermatocyte:	6.5pg
Secondary Oocyte:	6.5pg

Common errors:

  incorrect number of chromosomes

  incorrect chromosome morphology

    metacentric, acrocentric

  poor relationship of DNA content to cells

  inappropriate symbols

  inappropriate placement of genes

Question 3. Assume that you are examining a cell under a microscope and you observe the following as the total chromosomal constituents of a nucleus. You know that 2n=2 in this organism, that all chromosomes are telocentric, and that each G_1 cell nucleus contains 8 picograms of DNA.

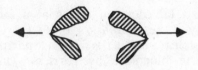

(a) Circle the correct stage for this cell:

anaphase of mitosis,

anaphase of meiosis I,

anaphase of meiosis II,

telophase of mitosis.

(b) How many picograms of chromosomal DNA would you expect in the cell shown above?

_____

> **Concepts:**
>
> **chromosome morphology**
>
> **telocentric, etc.**
>
> **anaphase configurations**
>
> **chromosome mechanics**
>
> **meiosis, mitosis**
>
> **DNA content in cell cycles**

Answer 3.

(a) Since the cell contains only two chromosomes (2n=2) and there are two chromosomes pictured, it can not represent a cell in the second phase (II) of meiosis. The chromosomes are telocentric which means that the centromere is at the end of the chromosome. When pulled at anaphase, two sideways "Vs" or "**< >**" would be expected and the cell would be at the anaphase stage of meiosis I.

The only other possibility to produce the "**< >**" figure would be a metaphase chromosome at anaphase of mitosis or anaphase II of meiosis. However these possibilities are negated because the chromosomes are stated as being telocentric.

(b) In order to get the correct answer for the second part one must consider that, because of the S-phase, at anaphase I the DNA complement is twice that of a G_1 cell. Therefore the correct answer is 16pg DNA.

> **Common errors:**
>
> **confusion on:**
>
> **significance of *telocentric***
>
> **significance of chromosome number**
>
> **many students consider the chromosomes to be metacentric**

Question 4. The human pedigree below indicates those individuals with a particular phenotype (shaded symbols). Males are represented by squares, females by circles.

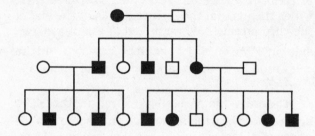

(a) Assume that this family is found in a location where approximately 65% of the people show the phenotype. Is the trait most likely caused by a dominant or recessive allele? Explain your reasoning.

(b) Assume that this family is found in a location where approximately .02% of the people show the phenotype. Is the trait most likely caused by a dominant or recessive allele? Explain your reasoning.

Concepts:

    Mendelian genetics

    dominant/recessive patterns

    pedigree analysis

    carrier status of some individuals

Answer 4. Looking at the pedigree by itself, nothing eliminates the possibility of the trait being caused by either a dominant allele or a recessive allele. The trait is seen in all generations which suggests that it is dominant, however, if the gene is recessive and "brought in" by the marriage partners (in the carrier state) then the gene could be recessive.

In part (a) the trait exists at a fairly high frequency in the population (65%), therefore it is very possible that all of the marriage partners from outside the "bloodline" are carriers. Under this condition homozygous recessive individuals would be seen at a relatively high frequency as in the pedigree.

Under condition (b) however, with the trait found relatively infrequently (0.02%), the likelihood of it being introduced from "carrier marriages" is quite low and a dominant mode of inheritance is favored.

Common errors:

    uncertainty of patterns of inheritance of dominant and recessive genes

    uncertainty of meaning of different locations having different gene frequencies

---

Question 5. The genes for *singed bristles* (*sn*) and *miniature wings* (*m*) are recessive and located on the X chromosome in *Drosophila melanogaster*. In a cross between a singed-bristled, miniature-winged female and a wild type male, all of the male offspring were singed-miniature.

(a) Draw meiotic metaphase I chromosomal configurations which represent the X and/or Y chromosomes of the parental (singed-miniature female and wild type male) flies. Place gene symbols (*sn*, *m*) and their wild type alleles (*sn*⁺, *m*⁺) on appropriate chromosomes.

(b) Draw a mitotic metaphase chromosomal configuration which represents the X and/or Y chromosomes of the F1 male. Place gene symbols (*sn*, *m*) on appropriate chromosomes.

(c) Most of the female offspring from the above-mentioned cross were phenotypically wild type; however one exceptional female was recovered which had singed bristles and miniature wings. Given that meiotic nondisjunction accounted for this exceptional female, would you expect it to have occurred in the parental male or parental female?

(d) Draw a meiotic, labeled (with gene symbols) circumstance and division product(s) which could account for the exceptional female described above. (Confine your drawing to X chromosomes only).

Concepts:

    chromosome mechanics

      meiosis

    symbolism

    DNA content (cell cycles)

    meiotic nondisjunction

    sex determination

Answer 5.

(a) The female parent would have the following labeled chromosomal symbolism remembering that at meiotic metaphase I, chromosomes are doubled, condensed, and synapsed.

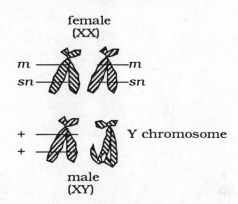

female
(XX)

male
(XY)

(b) In general, mitotic metaphase chromosomes are not synapsed, although in *Drosophila* mitotic chromosomes do pair. To avoid confusion and to be consistent with what is expected in other organisms, the mitotic chromosomes will not be drawn in the paired state.

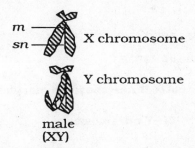

male
(XY)

(c) All of the female offspring from the above cross should be heterozygous and phenotypically wild type. The one exceptional female could have resulted from maternal nondisjunction at meiosis I or II, thus producing an egg cell with two X chromosomes, each containing the *sn* and *m* genes. When fertilized by a sperm cell carrying the Y chromosome (along with the normal haploid set of autosomes) an $X^{sn\,m}X^{sn\,m}Y$ female is produced.

(d)

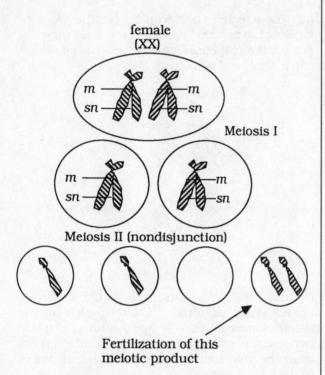

female
(XX)

m ——— *m*
sn ——— *sn*

Meiosis I

m ———
sn ———

———— *m*
———— *sn*

Meiosis II (nondisjunction)

Fertilization of this
meiotic product

Question 6. The red fox (*Vulpes vulpes*) has 17 pairs of somewhat long chromosomes. The Arctic fox (*Alopex lagopus*) has 26 pairs of somewhat shorter chromosomes.

(a) If a female red fox is crossed with a male Arctic fox, what will be the chromosome number in the somatic tissues of the hybrid?

(b) Assume that a somatic G_1 nucleus of the Arctic fox contains 12 picograms of DNA while a somatic G_1 nucleus of the red fox contains 8 picograms of DNA. How much nuclear DNA could you expect in a G_2 somatic nucleus of the hybrid?

Concepts:

> **meiosis and chromosome numbers**
>
> **DNA content**

Common errors:

> **incorrect chromosome morphology**
>
> **X, Y chromosmes**
>
> **inappropriate symbols**
>
> **inappropriate placement of genes**
>
> **problems with meiotic nondisjunciton**

Answer 6.

(a) Since the chromosome numbers are given in *pairs*, recall that during meiosis each gamete contains one chromosome of each pair. If the red fox has 17 pairs of chromosomes, then each gamete will contain 17 chromosomes. For the Arctic fox, each gamete should contain 26 chromosomes. A zygote is produced from the union of the parental gametes, therefore it should contain 43 chromosomes (17 + 26). It turns out that some such hybrids are viable but usually sterile because of developmental and chromosomal alignment and segregational problems at meiosis.

(b) If G_1 nuclei contain 12pg and 8pg DNA, then the gametes produced from these organisms will contain 6 and 4pg DNA respectively. Combining these gametes gives 10pg for a G_1 cell. For a G_2 cell there should be 20pg DNA.

Common errors:

confusion with *pairs* of chromosomes

gametic chromosome number

confusion with uneven number of chromosomes

confusion with *somatic* cells

Concepts:

sex-linked inheritance (X-linked)

pedigree construction

probability (product rule)

Question 7. Red-green colorblindness is inherited in man as an X-linked, recessive gene. Using the symbols below, draw a pedigree which is consistent with the following statements.

A phenotypically normal woman is married to a phenotypically normal man. The woman's parents are phenotypically normal but her maternal grandfather is colorblind. The woman's paternal grandparents as well as her maternal grandmother are phenotypically normal.

male	=	□
female	=	○
Rg	= **normal color sight**	
rg	= **colorblind**	

What is the probability that the first son born to the woman will be phenotypically normal (not be colorblind)?

Answer 7.

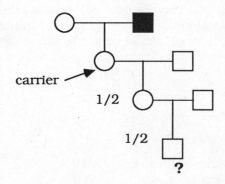

Because the maternal grandfather was colorblind, the woman's mother is a carrier for this X-linked gene ($X^{Rg}X^{rg}$). The woman therefore has a 1/2 chance of inheriting the X^{rg} chromosome from her mother and a 1/2 chance of passing this X^{rg} chromosome to her son. The chance that the son will receive the X^{rg} chromosome is therefore 1/4 (1/2 X 1/2). However, the question asks for the probability that the son will be normal.

The answer is therefore 1 minus 1/4 which equals 3/4.

Common errors:

> **difficulty in setting up pedigree**
>
> **inability to see independent probabilities**
>
> **multiplication of independent probabilities**
>
> **seeing that the *normal* is requested**

Concepts:

> **sex-linked inheritance (X-linked)**
>
> **chromosome mechanics**
>
> **meiosis**
>
> **conventional symbolism**

Question 8. In a *Drosophila* experiment a cross is made between a homozygous wild type female and a tan-bodied (mutant) male. All the resulting F_1 flies were phenotypically wild type. Adult flies of the F_2 generation (from a mating of the F_1's) had the following characteristics:

Sex	Phenotype	Number
Male	wild	346
Male	tan	329
Female	wild	702

(a) Using conventional symbolism, illustrate the genotype, *on an appropriate chromosomal configuration,* of a secondary oocyte nucleus of one of the F_1 females. Be certain to distinguish the X chromosomes from the autosomes. Note: *Drosophila melanogaster* has a diploid chromosome number of 8.

(b) Using the same conventional symbolism, give the genotype *on an appropriate chromosomal configuration* of a primary spermatocyte of the tan-bodied F_2 males.

Answer 8.

(a) First one must determine whether the gene for *tan body* is X-linked or autosomal (not on the sex chromosome). Because half of the F_2 males are mutant and half are wild type, and all the females are wild, the gene for *tan body* is behaving as X-linked. The F_1 female is heterozygous, therefore she should have either of the alleles (t, t^+) on the one X chromosome (a secondary oocyte has one representative of each chromosomal pair) in the following arrangement. Because *Drosophila* has 8 chromosomes, each secondary oocyte should have four chromosomes (including the X chromosome).

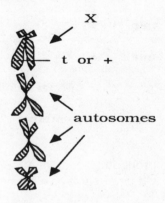

(b) A primary spermatocyte has the chromosomes in a doubled, condensed, and synapsed state. A tan-bodied male should have an X (containing a *t* gene) and Y chromosome as well as a diploid complement of autosomes.

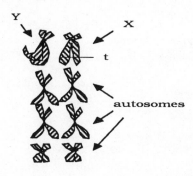

Common errors:

difficulty in recognizing X-linked inheritance

problems with placing genes on chromosomes

problems with visualizing genome

Question 9. *Gray* seed color in peas is dominant to *white*. Assume that Mendel conducted a series of experiments where plants were crossed and offspring classified according to the table below. What are the most probable genotypes of each parent?

Parents			Progeny	
---	---	---	gray	white
(a) gray	X	white	81	79
(b) gray	X	gray	120	42
(c) white	X	white	0	50
(d) gray	X	white	74	0

Concepts:

Mendelian genetics

monohybrid cross

dominance/recessiveness

3:1 and 1:1 ratios

Answer 9. First, assign gene symbols:

G = gray, gg = white

(a) Since there is an approximate 1:1 ratio in the progeny, the parental genotypes are *Gg* X *gg*.

(b) A 3:1 ratio is apparent, therefore the parental genotypes are *Gg* X *Gg*.

(c) Because there are no gray phenotypes and *gray* is the dominant allele, the parental genotypes must be *gg* X *gg*.

(d) Since there are no white types and the sample is sufficiently large, it is very likely that the parental genotypes are *GG* X *gg*.

Common errors:

students usually have only minor problems with this type of question

some careless, random mistakes

Question 10. Hemophilia (type A) is recessive and X-linked in humans whereas the ABO blood groups locus is autosomal. Assume that the following matings were examined for the transmission of these genes.   Give the expected phenotypes and numbers assuming 800 offspring are produced.

Group A: Females heterozygous for hemophilia with blood type AB mated to normal males with blood group O.

Group B: Females heterozygous for hemophilia with blood type AB mated to males with hemophilia and blood group AB.

Concepts:

   sex-linkage (X-linked)

   autosomal inheritance

   dihybrid situation

   incomplete dominance

   complete dominance

Answer 10. Set up the crosses with an appropriate symbol set such as the following:

   h = hemophilia

   H = normal allele

   $I^A I^B$ = AB blood group

   $I^o I^o$ = O blood group.

Group A.

male gametes

		HI^o	YI^o
	HI^A	$HHI^A I^o$	$HYI^A I^o$
	HI^B	$HHI^B I^o$	$HYI^B I^o$
female gametes	hI^A	$HhI^A I^o$	$hYI^A I^o$
	hI^B	$HhI^B I^o$	$hYI^B I^o$

Collecting phenotypes gives:

   1/4 female, normal, A blood (200)

   1/4 female, normal, B blood (200)

   1/8 male, normal, A blood (100)

   1/8 male, normal, B blood (100)

   1/8 male, hemophilia, A blood (100)

   1/8 male, hemophilia, B blood (100)

Group B. In this example the forked-line method will be used. Consider what will be happening for the *hemophilia* locus independently from the blood group locus.

1/4 females, normal ⟨ 1/4 A blood (50)
                      2/4 AB blood (100)
                      1/4 B blood (50)

1/4 females, hemophilia – 1/4 A blood (50)
                           2/4 AB blood (100)
                           1/4 B blood (50)

1/4 males, normal —— 1/4 A blood (50)
                     2/4 AB blood (100)
                     1/4 B blood (50)

1/4 males, hemophilia — 1/4 A blood (50)
                        2/4 AB blood (100)
                        1/4 B blood (50)

Concepts:

Mendelian patterns

1:1:1:1 ratio

null hypothesis

expected values

χ^2 analysis

interpretation of χ^2

Question 11. For the cross *PpRr X pprr* where complete dominance and independent assortment hold, assume that you received the following results and you wished to determine whether they differ significantly (in a statistical sense) from expectation.

PR phenotypes	=	40
Pr phenotypes	=	10
pR phenotypes	=	20
pr phenotypes	=	30

(a) State the null hypothesis associated with this test of significance.

(b) How many degrees of freedom would be associated with this test of significance?

(c) Assuming that a Chi-Square value of 20.00 is arrived at in this test of significance, do you accept or reject the null hypothesis?

Degrees of Freedom	P = 0.05
1	3.84
2	5.99
3	7.82
4	9.49
5	11.07

Answer 11.

(a) An appropriate null hypothesis for this example would be that the observed (measured) values do no not differ significantly from the predicted ratio of a 1:1:1:1. One might also say that any deviation between the observed and predicted values is due to chance and chance alone.

(b) Because there are four classes being compared, there will be three degrees of freedom.

(c) Given that the Chi-square value of 20.00 is considerably greater than 7.82 (for three degrees of freedom) the null hypothesis should be rejected and the conclusion be that the observed values differ significantly from the predicted values bases on a 1:1:1:1 ratio.

Question 12. Explain the processes, genotypic, chromosomal, and developmental which would lead to a bilateral gynandromorph in *Drosophila melanogaster* in which the male half of the fly has white eyes and singed bristles, while the female half is phenotypically wild type.

Concepts:

> **sex-linked inheritance (X-linked)**
>
> **gynandromorph production**
>
> **sex determination in *Drosophila***
>
> **insect development**
>
> **mitosis, nondisjunction**

Answer 12. In order for this type of fly to occur, the zygote must start out as a *cis* heterozygote in which both mutant genes are on one homologue and wild type alleles are on the other. In addition, one of the wild type chromosomes must get "lost" at the first mitotic division, thus making the female half $X^{++}X^{w sn}$ and the other half $X^{w sn}O$. the diagram below explains this situation.

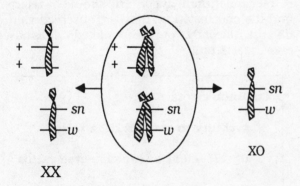

Because XX nuclei produce female tissue and XO nuclei produce male tissue, the phenotypes of the two sides are thus described. Developmentally, once the cleavage nuclei reach the peripheral areas of the egg to form a blastoderm, they become committed as to their adult fate. Because there is little "wandering" of nuclei either during their migration to the egg periphery or after they reach the periphery, the male/female boundary is quite clean.

Common errors:

> **difficulty in setting up the problem**
>
> **dealing with mitotic nondisjunction**
>
> **embryonic development of *Drosophila***

Question 13. Assume that there are 18 map units between two loci in the mouse and that you are able to microscopically observe meiotic chromosomes in this organism. If you examined 150 primary oocytes, in how many would you expect to see a chiasma between the two loci mentioned above?

Concepts:

> **crossing over mechanisms**
>
> **meiosis**
>
> **gene mapping**
>
> **chromosome mechanics**

Answer 13. The basis of the solution is to recall that crossing over occurs at the "four-strand stage" (after the S-phase) and each chiasma involves only two of the four chromatids present in each tetrad. Therefore for each chiasma only two of the four, or 1/2, of the chromatids are crossover chromatids.

Gene mapping basically is the process of dividing the number of crossover chromatids by the total number of chromatids. Since each chiasma involves only two of the four chromatids, the map distance must be half of the chiasma frequency. If there are 18 map units between two genes then the chiasma frequency would be 36%. If one examined 150 primary oocytes one would therefore expect to see 0.36 X 150, or 54 cells with a chiasma between the two loci.

Common errors:

problem seeing relationships

chiasma, map units

problems "seeing" meiosis

visualization of crossing over

Question 14. Given below are four dihybrid crosses between various strains of *Drosophila*. To the right of each are map distances known to exist between the genes involved. For each cross give the phenotypes of the offspring and the percentages (in the parentheses) expected for each.

	Mating		
	Female	*Male*	*Map distance*
(a)	*AB/ab*	*ab/ab*	20
(b)	*Pq/pQ*	*pq/pq*	50
(c)	*DB/db*	*db/db*	0
(d)	*ab/ab*	*AB/ab*	20

Concepts:

linkage and crossing over

computation of map units

complete linkage

independent assortment

lack of crossing over in males

Answer 14. The key to solving these type of "reverse mapping" problems is to keep in mind that a map unit is computed by the equation

$$\frac{\text{Number of crossover progeny}}{\text{Total number of progeny}} \text{ X } 100$$

and that if the map distance is given it is easy to determine the percentages of parental and crossover offspring. Remember that there are two classes of crossovers and two classes of parentals from each cross.

(a)

AB/ab = 40%, ab/ab = 40% (parentals)

Ab/ab = 10%, aB/ab = 10% (crossovers)

(b)

Pq/pq = 25%, pQ/pq = 25% (parentals)

PQ/pq = 25%, pq/pq = 25% (crossovers)

Notice that this is independent assortment.

(c) DB/db = 50%, db/db = 50% (all parentals)

(d) AB/ab = 50%, ab/ab = 50% (all parentals)

The reason that there are all parentals and no crossovers in this cross is that there is no crossing over in male *Drosophila.* With no crossing over, the *AB/ab* chromosomes in the male are passed to gametes without crossovers.

Common errors:

> **difficulty in going from map units to frequencies of classes of offspring**

> **failure to see that there are two parental and two crossover classes**

> **minor, careless mistakes**

Concepts:

> **mapping relative to the centromere**

> **working from map units to expected classes**

> **first and second division segregation**

Answer 15. The key to answering this problem is to remember that the map distance from a gene to the centromere is calculated by dividing the percentage of second division asci by two.

+	*f*	+	+	*f*
f	+	*f*	+	*f*
+	*f*	*f*	*f*	+
f	±	±	*f*	±

The second division asci are those in which a crossover has occurred between the gene in question (*f*) and the centromere as shown below. Such second division asci are boxed above.

Question 15. Lindegren (1933) determined that the *fluffy* locus in *Neurospora* is approximately 30 map units from the centromere. If you examined 100 asci, in what actual numbers would you expect to see the following asci classes? Place expected numbers in the spaces below the asci; they should add up to 100. Note: there are numerous correct answers possible.

+	*f*	+	+	*f*
f	+	*f*	+	*f*
+	*f*	*f*	*f*	+
f	±	±	*f*	±

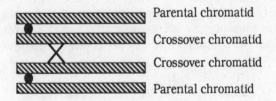

Parental chromatid
Crossover chromatid
Crossover chromatid
Parental chromatid

To answer the problem one would apply numbers so that there are 60% second division asci and 40% first division asci. Thirty map units will result when the value of 60% for the second division asci is divided by two.   One example is given below.

```
+  ƒ  +  +  ƒ
ƒ  +  ƒ  +  ƒ
+  ƒ  ƒ  ƒ  +
ƒ  ±  ±  ƒ  ±
20 20 20 20 20
```

Common errors:

   difficulty in working mapping problems

   chiasma frequency is 2X map units

   simple mathematics and percentages

   recognition of first and second division arrangements

DNA - The Genetic Material

Vocabulary: Organization and Listing of Terms

Historical

1900-1940

   genetic material

      proteins

      nucleic acids

         tetranucleotide hypothesis

         base ratios

            Erwin Chargaff

         transforming principle

            Avery *et al.* (1944)

        T2 bacteriophage (phage)

            Hershey and Chase (1952)
            ^{32}P, ^{35}S

Structures and Substances

Messenger RNA (mRNA)

Transfer RNA (tRNA)

Ribosomal RNA (rRNA)

   ribosome

Ribonuclease

Deoxyribonuclease

Lysozyme

Protoplasts (spheroplasts)

Insulin

Interferon

Human ß-globin gene

RNA core

Coat protein

Qß RNA replicase

Reverse Transcriptase

Processes/Methods

Replication (F8.1)

    mitosis

    meiosis

Storage of information(F8.1)

Expression(F8.1)

    transcription

    translation

Variation (mutation)(F8.1)

Transformation

    Diplococcus pneumoniae

        virulent

        avirulent

        serotypes (II, III)

        heat-killed IIIS

        rough, extremely rough (ER)

            ribonuclease

            deoxyribonuclease

    transfection

    recombinant DNA research

        insulin

        interferon

        human ß-globin gene

            transgenic mice

                human growth hormone

RNA as genetic material

    TMV (tobacco mosaic virus)

    Holmes ribgrass (HR)

        RNA core

        coat protein

    Qß phage

        Qß RNA replicase

    retroviruses

        reverse transcription

Concepts

Storage of genetic information

Information flow

Transformation

Differential labeling of macromolecules

Circumstantial evidence

    DNA distributions

    mutagenesis

        action spectrum

        absorption spectrum

            260 nm, 280 nm

F8.1. Illustration of relationships between DNA, its functions, and related products.

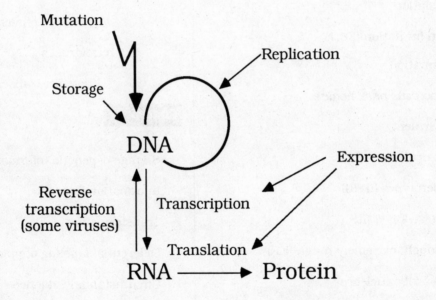

Solutions to Problems and Discussion Questions

1. *Replication* is that process which leads to the production of identical copies of the existing genetic information. Since daughter cells contain essentially exact copies (with some exceptions) of genetic information of the parent cell, and through the production and union of gametes, offspring contain copies (with variation) of parental genetic information, the genetic material must make copies (replicate) of itself. Replication is accomplished during the S phase of interphase.

The genetic material is capable of *expression* through the production of a phenotype. Through transcription and translation, proteins are produced which contribute to the phenotype of the organism. The genetic material must be stable enough to maintain information in "*storage*" from one cell to the next and one organism to the next. Because the genetic material is not "used up" in the processes of transcription and translation, genetic information can be stored and used constantly.

Above, it was stated that the genetic material must be stable enough to store genetic information; however, variation through *mutation* provides the raw material for evolution. The genetic material is capable of a variety of changes, both at the chromosomal (see Chapter 7) and nucleotide levels. See F8.1.

2. Prior to 1940 most of the interest in genetics centered on the transmission of similarity and variation from parents to offspring (transmission genetics). While some experiments examined the possible nature of the hereditary material, abundant knowledge of the structural and enzymatic properties of proteins generated a bias which worked to favor proteins as the hereditary substance. In addition, proteins were composed of as many as twenty different subunits (amino acids) thereby providing ample structural and functional variation for the multiple tasks which must be accomplished by the genetic material. The tetranu-cleotide hypothesis (structure) provided insufficient variability to account for the diverse roles of the genetic material.

3. Griffith performed experiments with different strains of *Diplococcus pneumoniae* in which a heat-killed pathogen, when injected into a mouse with a live non-pathogenic strain, eventually led to the mouse's death. A summary of this experiment is provided in K/C-Fig.8.3. Examination of the dead mice revealed living pathogenic bacteria. Griffith suggested that the heat-killed virulent (pathogenic) bacteria transformed the avirulent (non-pathogenic) strain into a virulent strain.

Avery and coworkers systematically searched for the transforming principle originating from the heat-killed pathogenic strain and determined it to be DNA. Taylor showed that transformed bacteria are capable of serving as donors of transforming DNA indicating that the process of transformation involves a stable alteration in the genetic material (DNA).

4. Transformation is dependent on a macromolecule (DNA) which can be extracted and purified from bacteria. During such purification however, other macromolecular species may contaminate the DNA. Specific degradative enzymes, proteases, RNase, and DNase were used to selectively eliminate components of the extract and if transformation is concomitantly eliminated, then the eliminated fraction is the transforming principle. DNase eliminates DNA and transformation, therefore it must be the transforming principle.

5. On the surface, transformation seems like a relatively simple process of a bacterial cell incorporating exogenous DNA into its own DNA. However, what are the actual parameters and mechanisms which influence transformation?

Are all bacterial species capable of being transformed?

Are all bacteria within a given culture transformed with equal frequency?

Are there certain phases of the bacterial growth cycle which contain more "competent" bacteria than other phases?

Is single-stranded DNA as capable of transformation as double-stranded DNA?

Is linear DNA as efficient in transformation as circular DNA?

How much DNA is taken up by a recipient bacterium and what percentage of the DNA is genetically active?

Is there a limit to the number of molecules of DNA that a single bacterial cell can take in?

What is the relationship between incoming DNA and DNA of the host?

6. Nucleic acids contain large amounts of phosphorus and no sulfur whereas proteins contain sulfur and no phosphorus. Therefore the radioisotopes ^{32}P and ^{35}S will selectively label nucleic acids and proteins, respectively.

The Hershey and Chase experiment is based on the premise that the substance injected into the bacterium is the substance responsible for producing the progeny phage and therefore must be the hereditary material. The experiment demonstrated that most of the ^{32}P-labeled material (DNA) was injected while the phage ghosts (protein coats) remained outside the bacterium. Therefore the nucleic acid must be the genetic material.

7. Actually phosphorus is found in approximately equal amounts in DNA and RNA. Therefore labeling with ^{32}P would "tag" both RNA and DNA. However, the T2 phage, in its mature state, contains very little if any RNA, therefore DNA would be interpreted as being the genetic material in T2 phage.

8. In theory, the general design would be appropriate in that some substance, if labeled, would show up in the progeny of transformed bacteria. However, since the amount of transforming DNA is extremely small compared to the genomic DNA of the recipient bacterium and its progeny, it would be technically difficult to assay for the labeled nucleic acid. In addition, it would be necessary to know that the small stretch of DNA which caused the genetic transformation was actually labeled. This in itself would be relatively easy using present-day recombinant DNA techniques; however, in earlier times, such specific labeling would have been difficult.

9. The early evidence would be considered circumstantial in that at no time was there an experiment, like transformation in bacteria, in which genetic information in one organism was transferred to another using DNA. Rather, by comparing DNA content in various cell types (sperm and somatic cells) and observing that the *action* and *absorption* spectra of ultraviolet light were correlated, DNA was considered to be the genetic material. This suggestion was supported by the fact that DNA was shown to be the genetic in bacteria and some phage. Direct evidence for DNA being the genetic material comes from a variety of observations including gene transfer which has been facilitated by recombinant DNA techniques.

10. Some viruses contain a genetic material composed of RNA. The tobacco mosaic virus is composed of an RNA core and a protein coat. "Crosses" can be made in which the protein coat and RNA of TMV are interchanged with another strain (Holmes ribgrass). The source of the RNA determines the type of lesion, thus, RNA is the genetic material in these viruses. Retroviruses contain RNA as the genetic material and use an enzyme known as *reverse transcriptase* to produce DNA which can be integrated into the host chromosome. See F8.1.

Nucleic Acids - Structure and Analysis

Vocabulary: Organization and Listing of Terms

Historical

Watson and Crick (1953)

Chargaff

Franklin, Wilkins

Structures and Substances

Nucleic acids

  nucleotides

  mononucleotides

    nitrogenous base

      purines

        adenine

        guanine

      pyrimidines

        cytosine

        thymine

        uracil

pentose sugar

  ribose

  deoxyribose

  phosphoric acid

nucleoside

  monophosphate

  diphosphate

  triphosphate

    adenosine triphosphate

inorganic phosphate

  hydrolysis

polynucleotide

  phosphodiester bond

    C-5' to C-3'

double helix

base composition

  A=T, G=C

  (A+G) = (C+T)

  (A+T)/(C+G) = variable

X-ray diffraction

plectonic

antiparallel

0.34 nm (stacked bases)

3.4 nm (complete turn)

10 bases per turn

  10.4 bases per turn

hydrogen bonds

  (A to T, G to C)

complementarity

major and minor grooves

20 nm diameter

hydrophobic bases

hydrophilic backbone

A-DNA

B-DNA

C-DNA

D-DNA

E-DNA

Z-DNA

left-handed double helix

zigzag conformation

dimensions

major and minor grooves

RNA

  flexibility

    ribosomal RNA (rRNA)

      ribosomes

    messenger RNA (mRNA)

    transfer RNA (tRNA)

Processes/Methods

Bonding

  sugar to purine

  sugar to pyrimidine

  nucleotide to nucleotide

Single crystal X-ray analysis

Svedberg coefficient (S)

Absorption of ultraviolet light (UV)

  254-260 nm

Paper chromatography

Sedimentation behavior

  sedimentation velocity

  sedimentation equilibrium

    (density gradient)

Spectrophotometry

Denaturation (melting)

   heat

   chemical treatment

Melting profile

Melting temperature (T_m)

Hyperchromic shift

Renaturation (hybridization)

   DNA/DNA

   DNA/RNA

   in situ hybridization

   kinetics

   $C_o t$

   $C_o t_{1/2}$

   complexity (X)

      repetitive DNA

      moderately (intermediate)

         repetitive DNA

highly repetitive

   (simple sequence) DNA

single copy (unique sequence) DNA

short-period interspersion

long-period interspersion

Concepts

Genetic variation

Model building

   DNA double helix

   semiconservative replication

   storage of genetic information

   mutation

Genomic complexity

   reassociation kinetics

Solutions to Problems and Discussion Questions

1. The structure of deoxyadenylic acid is given below and in K/C-Fig.9.3. Linkages among the three components require the removal of water (H_2O).

2. The numbering of the carbons on the sugar is especially important (see diagram below). Examine K/C-Figs.9.1 and 9.2 for the numbers on the carbons and nitrogens of the bases:

3. Examine the structures of the bases in K/C-Fig.9.1. The other bases would be named as follows:

Guanine: 2-amino-6-oxypurine

Cytosine: 2-oxy-4-aminopyrimidine

Thymine: 2,4-dioxy-5-methylpyrimidine

Uracil: 2,4-dioxypyrimidine

4. Examine K/C-Fig.9.4 for the format for this drawing. Note that the complementary strand must be drawn in the antiparallel orientation.

5. The following are characteristics of the Watson-Crick double-helix model for DNA:

The base composition is such that A=T, G=C and (A+G) = (C+T). Bases are stacked, 0.34 nm apart, in a plectonic, antiparallel manner. There is one complete turn for each 3.4 nm which constitutes 10 bases per turn. Hydrogen bonds hold the two polynucleotide chains together, each being formed by phosphodiester linkages between the sugars and the phosphates. There are two hydrogen bonds forming the A to T pair and three forming the G to C pair. The double helix exists as a twisted structure, approximately 20 nm in diameter, with a topography of major and minor grooves. The hydrophobic bases are located in the center of the molecule while the hydrophilic phosphodiester backbone is on the outside.

6. In addition to creative "genius" and persever-ance, model building skills, and the conviction that the structure would turn out to be "simple" and have a natural beauty in its simplicity, Wat-son and Crick employed the X-ray diffraction information of Franklin and Wilkins, and the base ratio information of Chargaff.

7. Because in double-stranded DNA, A=T and G=C (within limits of experimental error), the data presented would have indicated a lack of pairing of these bases in favor of a single-stranded struc-ture or some other nonhydrogen-bonded struc-ture.

Alternatively, from the data it would appear that A=G and T=C which would require purines to pair with purines and pyrimidines to pair with py-rimidines. In that case, the DNA would have contradicted the data from Franklin and Watkins which called for a constant diameter for the double-stranded structure.

8. A covalent bond is a relatively strong bond which involves the sharing of electrons between two or more atoms. Hydrogen bonds, much weaker than covalent bonds, are formed as a result of

"electrostatic attraction between a co-valently bonded hydrogen atom and an atom with an unshared electron pair. The hydrogen atom assumes a partial positive charge, while the unshared electron pair, characteristic of covalently bonded oxygen and nitrogen atoms, assumes a partial negative charge. These opposite charges are responsible for the weak chemical at-traction." (Klug and Cummings)

Complementarity, responsible for the chemi-cal attraction between adenine and thymine (uracil) and guanine and cytosine, is responsible for DNA and RNA assuming their double-stranded charac-ter. Complementarity is based on hydrogen bond-ing.

9. Three main differences between RNA and DNA are the following:

(1) uracil in RNA replaces thymine in DNA,

(2) ribose in RNA replaces deoxyribose in DNA, and

(3) RNA often occurs as both single- and double-stranded forms whereas DNA most often occurs in a double-stranded form.

10. While there are many types of RNA, the three main types described in this section are presented below:

ribosomal RNA: rRNA combines with proteins to form ribosomes which function to align mRNA and charged tRNA molecules during translation.

transfer RNA: tRNAs are involved in protein syn-thesis in that they represent a "link" between the codes in DNA (as reflected in mRNA) and the ordering of amino acids in proteins. Transfer RNAs are specific in that each species is attached to only one type of amino acid.

messenger RNA: the genetic code in DNA is transferred to the site of protein synthesis by a relatively short-lived molecule called messenger RNA. In eukaryotes, mRNA carries genetic infor-mation from the nucleus to the cytoplasm. It is the sequence of bases in mRNA which specifies the order of amino acids in proteins.

11. The nitrogenous bases of nucleic acids (nu-cleosides, nucleotides, and single- and double-stranded polynucleotides), absorb UV light maximally at wavelengths 254 to 260 nm. Using this phenomenon, one can often determine the presence and concentration of nucleic acids in a mixture. Since proteins absorb UV light maxi-mally at 280 nm, this is a relatively simple way of dealing with mixtures of biologically important molecules.

UV absorption is greater in single-stranded molecules (hyperchromic shift) as compared to double-stranded structures, therefore one can easily determine, by applying denaturing conditions, whether a nucleic acid is in the single- or double-stranded form. In addition, A-T rich DNA denatures more readily than G-C rich DNA, therefore one can estimate base content by denaturation kinetics.

12. *Sedimentation velocity* centrifugation refers to an ultracentrifugation technique which monitors the velocity with which macromolecules move through a centrifugal field. Molecules move through the gradient on the basis of their mass and shape. If centrifuged long enough, such molecules will end up at the bottom of the tube.

Sedimentation equilibrium centrifugation is a technique which provides separation in a gradient on the basis of buoyant density. Macromolecules migrate through the gradient until they reach and subsequently remain at the point of equal density.

13. The mass and shape of a molecule will be influential in determining a molecule's Svedberg coefficient. Since shape of a molecule is influenced by a variety of factors, such as *pH*, degree of hydration, and temperature, the environment of the molecule must he carefully standardized.

14. Guanine and cytosine are held together by three hydrogen bonds whereas adenine and thymine are held together by two. Because G-C base pairs are more compact, they are more dense than A-T pairs. The percentage of G-C pairs in DNA is thus proportional to the buoyant density of the molecule as illustrated in K/C-Fig.9.13.

15. Various treatments, heat, and certain chemical environments, cause separation of the hydrogen bonds which hold together the complementary strands of DNA. Under these conditions, double-stranded DNA is changed to single-stranded DNA.

16. A *hyperchromic effect* is the increased absorption of UV light as double-stranded DNA (or RNA for that matter) is converted to single-stranded DNA. As illustrated in K/C-Fig.9.14, the change in absorption is quite significant, with a structure of higher G-C content *melting* at a higher temperature than an A-T rich nucleic acid. If one monitors the UV absorption with a spectrophotometer during the melting process the hyperchromic shift can be observed. The T_m is the point on the profile (temperature) at which half (50%) of the sample is denatured.

17. Because G-C base pairs are formed with three hydrogen bonds while A-T base pairs by two such bonds, it takes more energy (higher temperature) to separate G-C pairs.

18. *Denaturation* is similar to *melting*, both meaning the separation of double-stranded structures. *Renaturation* and *hybridization* are similar, both referring to the reassociation of separated strands.

See K/C-Fig.9.15.

19. The reassociation of separate complementary strands of a nucleic acid, either DNA or RNA, is based on hydrogen bonds forming between A-T (or U) and G-C.

20. In one sentence of Watson and Crick's paper in *Nature,* they state "It has not escaped our notice that the specific pairing we have postulated immediately suggests a possible copying mechanism for the genetic material." The model itself indicates that unwinding of the helix and separation of the double-stranded structure into two single strands immediately exposes the specific hydrogen bonds through which new bases are brought into place.

21. Carefully examine K/C-Figs.9.15, 9.17, 9.18, and 9.20.   First understand the concept of molecular hybridization, then see that as the degree of strand uniqueness increases, the time required for reassociation increases. Repetitive sequences renature relatively quickly because the likelihood of complementary strands interacting increases.

For curve *A* in the problem, there is evidence for three different species of molecules.   The fraction which reassociates faster than the *E. coli* DNA is highly repetitive, the second "hump" is moderately repetitive, and the last fraction (with the highest $C_o t_{1/2}$ value) contains primarily unique sequences.   Fraction *B* contains mostly unique, relatively complex DNA.

22. As mentioned in the previous problem such a mixture of the DNAs would give a "$C_o t_{1/2}$ -curve" like that in *A* .

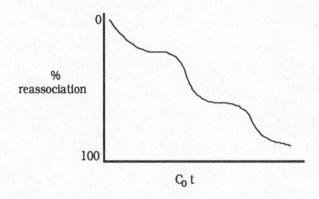

23.

(1) As shown, the extra phosphate is not normally expected.

(2) In the adenine ring, a nitrogen is at position 8 rather than position 9.

(3) The bond from the C'-1 to the sugar should form with the N at position 9 (N-9) of the adenine.

(4) The dinucleotide is a "deoxy" form, therefore each C-2' should not have a hydroxyl group. Notice the hydroxyl group at C'-2 on the sugar of the adenylic acid.

(5) At the C-5 position on the thymine residue, there should be a methyl group.

(6) At the C'-5 position on the thymidylic acid, there is an extra OH group.

24.   The general process is one of semiconservative replication, however, the bottom portion, the single double-stranded region, coils to the left, whereas the two top strands coil to the right.

25. As provided in the text and especially in the *Insights and Solutions* section, a direct proportionality between $C_o t_{1/2}$ and the number of base pairs exists under certain conditions.

The ratios for MS-2 would be as follows:

$$0.5/10^5 = 0.001/ X$$

or

$$X/0.001 = 10^5/0.5$$

$$X = (0.001)(10^5)/0.5$$

$$X = 200 \text{ base pairs}$$

The ratios for *E. coli* would be as follows:

$$0.5/10^5 = 10.0/ X$$

or

$$X/10.0 = 10^5/0.5$$

$$X = (10.0)(10^5)/0.5$$

$$X = 2 \times 10^6 \text{ base pairs}$$

26. Without knowing the exact bonding characteristics of hypoxanthine or xanthine it may be difficult to predict the likelihood of each pairing type. It is likely that both are of the same class (purine or pyrimidine) because the names of the molecules indicate a similarity. In addition, the diameter of the structure is constant which, under the model to follow, would be expected.   In fact, hypoxanthine and xanthine are both purines.

Because there are equal amounts of A, T and H, one could suggest that they are hydrogen bonded to each other; the same may be said for C, G, and X. Given the molar equivalence of erythrose and phosphate, an alternating sugar-phosphate-sugar backbone as in "earth-type" DNA would be acceptable. A model of a triple helix would be acceptable, since the diameter is constant. Given the chemical similarities to "earth-type" DNA it is probable that the unique creature's DNA follows the same structural plan.

10

DNA Replication, Synthesis, and Recombination

Vocabulary: Organization and Listing of Terms

Structures and Substances

DNA polymerase I

  polA1

Spleen phosphodiesterase (F10.1)

5'-nucleotides (F10.2)

3'-nucleotides (F10.2)

Phage øX174

  + strand

  - strand

  replicative form (RF)

  5-bromodeoxyuridine

DNA ligase (polynucleotide joining enzyme)

DNA polymerase II

DNA polymerase III

  holoenzyme

  subunits

dimer

  replisome

Helicase

Single-stranded DNA binding proteins

DNA gyrase

RNA primer

  primase

  free hydroxyl group

DNA ligase

  ligase deficient mutant

ori C

  9mer, 13mer

dnaA, dnaB, dnaC

Eukaryotic DNA polymerases

  four kinds

  semidiscontinuous

Euchromatin, heterochromatin

nucleosome

Endonuclease

Heteroduplex DNA molecules

Holliday structures

chi form

recombinant duplexes

recA

Processes/Methods

Replication of DNA

semiconservative

Meselson and Stahl - 1958

E. coli

equilibrium sedimentation

^{15}N, ^{14}N

Taylor, Woods, and Hughes - 1957

Vicia faba

^{3}H-thymidine

autoradiography

sister chromatid exchanges

conservative

dispersive

bidirectional (vs. unidirectional)

origin of replication, *ori*

replicon

multiple origins (not random)

replication fork

continuous, discontinuous

leading strand

lagging strand

Okazaki fragments

replication eye (bubble)

rolling circle model

unit genome

concatemer

Synthesis of DNA *in vitro*

reaction mixture

nearest neighbor frequency test (F10.1)

spleen phosphodiesterase

biologically active DNA

transfection of *E. coli*

faithful copying

Genetic recombination

single-stranded nick

endonuclease

Gene conversion

Neurospora

nonreciprocal

Concepts

Replication

semiconservative

antiparallel

continuous, discontinuous

conservative

dispersive

rolling circle model

unit genome

Conditional mutants (F10.3)

F10.1. Illustration of the mode of action of spleen phosphodiesterase.

Cleavage with
spleen phosphodiesterase

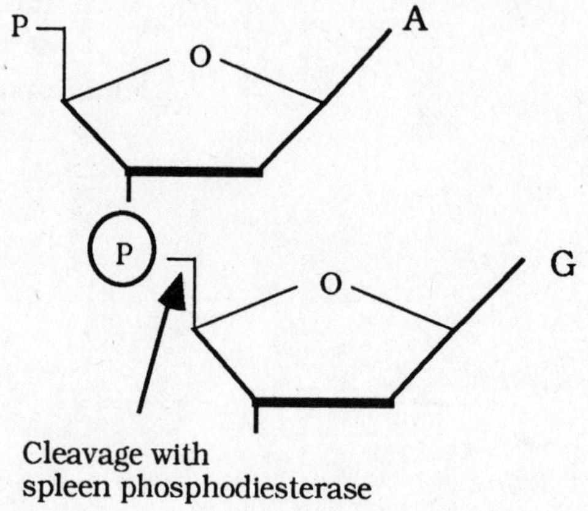

F10.2 Shorthand structures for 3' and 5' nucleotides.

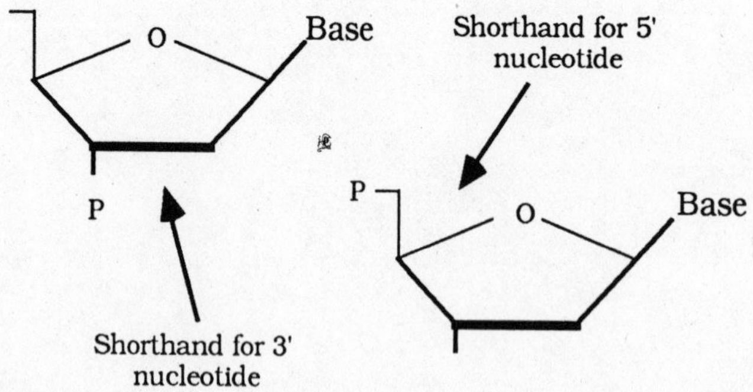

Shorthand for 3'
nucleotide

Shorthand for 5'
nucleotide

F10.3. Illustration of the influence of a conditional mutant on protein structure and therefore function.

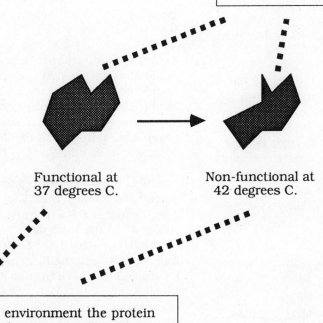

Changes in the active site(s) or functional domain(s) of a protein will alter function of that protein.

Functional at 37 degrees C.

Non-functional at 42 degrees C.

Changes in the environment the protein may cause conformational changes in theprotein.

Solutions to Problems and Discussion Questions

1. Refer to K/C-Figs.10.2 and 10.4 as well as Question #1 in the *Insights and Solutions* section. The differences among the three models of DNA replication relate to the manner in which the new strands of DNA are oriented as daughter DNA molecules are produced.

Conservative: In the conservative scheme, the original daughter strand remains as a complete unit and the new DNA double helix is produced as a single unit. The old DNA is completely *conserved.*

Semiconservative: Each daughter strand is composed of one old DNA strand and one new DNA strand. Separation of hydrogen bonds is required.

Dispersive: In the dispersive scheme, the original DNA strand is broken into pieces and the new DNA in the daughter strand is interspersed among the old pieces. Separation of covalent (phosphodiester) bonds is required for this mode of replication.

2. The Meselson and Stahl experiment has the following components.

(a) By labeling of the nitrogenous bases of the DNA of *E. coli* with a heavy isotope of ^{15}N, it would be possible to "follow" the "old" DNA. This is accomplished by growing the cells for many generations in medium containing ^{15}N.

(b) Cells were transferred to ^{14}N medium so that "new" DNA could be detected.

(c) A comparison of the density of DNA samples at various times in the experiment (initial ^{15}N culture, and subsequent cultures grown in the ^{14}N medium), showed that after one round of replication in the ^{14}N medium, the DNA was half as dense (intermediate) as the DNA from bacteria grown only in the ^{15}N medium.

In a sample taken after two rounds of replication in the ^{14}N medium, half of the DNA was of the intermediate density and the other half was as dense as DNA containing only ^{14}N DNA.

(d) Since the semiconservative model of replication states that each daughter strand of DNA is "semiconserved" the results stated above support the model.

3. Under a conservative scheme the first round of replication in ^{14}N medium produces one dense double helix and one "light" double helix in contrast to the intermediate density of the DNA in the semiconservative mode. Therefore, after one round or replication in the ^{14}N medium the conservative scheme can be ruled out.

After one round of replication in ^{14}N under a dispersive model, the DNA is of intermediate density, just as it is in the semiconservative model. However, in the next round of replication in ^{14}N medium, the density of the DNA is between the intermediate and "light" densities. Refer to K/C-Fig.10.2 if you have trouble answering this question.

4. Refer to K/C-Fig.10.5 for an illustration of the labeling of *Vicia* chromosomes under a Taylor, Woods, and Hughes experimental design. Notice that only those cells which pass through the S phase in the presence of the ^{3}H-thymidine are labeled and that each double helix (per chromatid) is "half-labeled."

(a) Under a conservative scheme all of the newly labeled DNA will go to one sister chromatid, while the other sister chromatid will remain unlabeled. In contrast to a semiconservative scheme, the first replicative round would produce one sister chromatid which has label on both strands of the double helix.

Conservative Replication

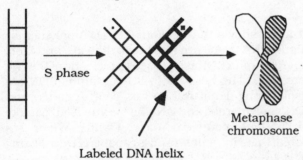

S phase

Labeled DNA helix

Metaphase chromosome

Dispersive Replication

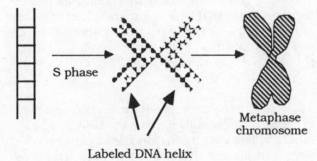

S phase

Labeled DNA helix

Metaphase chromosome

(b) Under a dispersive scheme all of the newly labeled DNA will be interspersed with unlabeled DNA. Because these preparations (metaphase chromosomes) are highly coiled and condensed structures derived from the "spread out" form at interphase (which includes the S phase) it is impossible to detect the areas where label is not found. Rather, both sister chromatids would appear as evenly labeled structures.

5. A triple helix was proposed in Problem 26 in Chapter 9. Because the semiconservative scheme predicts that *half* of the DNA in each daughter double helix is labeled, it would be difficult to envision a scheme where three strands are replicated in such a semiconservative manner. It would seem that either the conservative or dispersive scheme would fit more appropriately. To examine the nature of replication, one could devise an experiment similar to that of Meselson and Stahl or Taylor, Woods, and Hughes.

6. The *in vitro* replication requires a DNA template, a divalent cation (Mg^{++}), and all four of the deoxyribonucleoside triphosphates: dATP, dCTP, dTTP, and dGTP. The lower case "d" refers to the deoxyribose sugar.

7. Prior to the development of highly efficient methods of enzyme isolation, large cultures, containing large numbers of bacterial cells, were needed to yield even small quantities of enzymes.

8. Two general analytical approaches showed that the products of DNA polymerase I were probably *copies* of the template DNA. Because *base composition* can be similar without reflecting sequence similarity, the least stringent test was the comparison of base composition. By comparing *nearest neighbor frequencies*, Kornberg determined that there is a very high likelihood that the product was of the same base sequence as the template.

9. Because base composition can be similar without reflecting sequence similarity, the least stringent test was the comparison of base composition. By comparing nearest neighbor frequencies Kornberg determined that there is a very high likelihood that the product was of the same base sequence as the template. However, since nearest neighbor analysis relates "neighbor frequency" and not base sequence, there is a chance that sequence differences are not detected.

10. The *in vitro* rate of DNA synthesis using DNA polymerase I is slow, being more effective at replicating single-stranded DNA than double-stranded DNA. In addition it is capable of degrading as well as synthesize DNA. Such degradation suggested that it functioned as a repair enzyme. In addition, DeLucia and Cairns discovered a strain of *E. coli* (*polA1*) which still replicated its DNA but was deficient in DNA polymerase I activity.

11. The overall plan of the *nearest neighbor experiment* is to determine the frequency of neighbors of a given base, say adenine, in both the template DNA and the product DNA. Should the frequency of neighbors be similar, then the sequence of bases is likely to be the same. This argument is strengthened if the frequencies of neighbors of the other bases (G, C, and T) are also similar in the template and primer.

The procedure is to introduce labeled phosphate by use of labeled (innermost phosphate) 5'-nucleotides as shown in K/C-Fig.10.8. In the example, cytidine has a ³²P attached to the C-5' atom.   Spleen phosphodiesterase cleaves the polymer at the C-5' position (between the C-5' and the phosphate) thereby transferring the labeled phosphate to the C-3' atom of its 5' neighbor.

Chromatographic separation of the resulting 3'-nucleotides, followed by determination of the radioactivity of each, provides an estimate of the frequency of which each base is a neighbor of cytosine. By repeating this procedure using other bases and other templates, one can be somewhat confident that the template is similar in base sequence to the *in vitro* product.

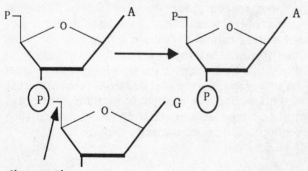

Cleavage with
spleen phosphodiesterase

12. As stated in the text, *biologically active* DNA implies that the DNA is capable of supporting typical metabolic activities of the cell or organism and is capable of faithful reproduction.

13. ØX174 is a well-studied single-stranded virus (phage) which can be easily isolated. It has a relatively small DNA genome (5500 nucleotides) which, if mutated, usually alters its reproductive cycle.

14. As shown in K/C-Fig.10.9, DNA polymerase I and DNA ligase are used to synthesize and label an RF duplex. DNase is used to nick one of the two strands. The heavy (³²P/BU-containing) DNA is isolated by denaturation and centrifugation and DNA polymerase and DNA ligase are used to make a synthetic complementary strand. When isolated, this synthetic strand is capable of transfecting *E. coli* protoplasts, from which new ØX174 phages are produced.

15. The *polA1* mutation was instrumental in demonstrating that DNA polymerase I activity was not necessary for the *in vivo* replication of the *E. coli* chromosome. Such an observation opened the door for the discovery of other enzymes involved in DNA replication.

16. All three enzymes share several common properties. First none can *initiate* DNA synthesis on a template but all can *elongate* an existing DNA strand assuming there is a template strand as shown in the figure below. Polymerization of nucleotides occurs in the 5' to 3' direction where each 5' phosphate is added to the 3' end of the growing polynucleotide. All three enzymes are large complex proteins with a molecular weight in excess of 100,000 daltons and each has 3' to 5' exonuclease activity.

Synthesis of DNA
can be initiated here

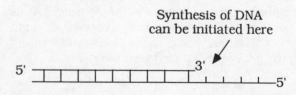

DNA polymerase I:
  5' to 3' exonuclease activity
  present in large amounts
  relatively stable
  removal of RNA primer

DNA polymerase II:
  function uncertain

DNA polymerase III:
  5' to 3' exonuclease activity
  relatively unstable
  essential for replication
  complex molecule
      seven polypeptide chains
      possibly 750,000 daltons

17. Given a stretch of double-stranded DNA, one could initiate synthesis at a given point and either replicate strands in one direction only (unidirectional) or in both directions (bidirectional) as shown below. Notice that in K/C-Fig. 10.12 the synthesis of complementary strands occur in a *continuous* 5' > 3' mode on the leading strand in the direction of the replication fork, and in a *discontinuous* 5' > 3' mode on the lagging strand opposite the direction of the replication fork. Such discontinuous replication forms Okazaki fragments.

Unidirectional model

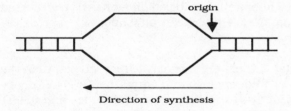

Direction of synthesis

Bidirectional model

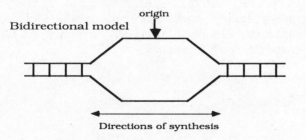

Directions of synthesis

18.   *Helicase, dnaA* and *single-stranded DNA binding* proteins initially unwind, open, and stabilize DNA at the initiation point. *DNA gyrase*, a DNA topoisomerase, relieves supercoiling generated by helix unwinding. This process involves breaking both strands of the DNA helix.

19. *Okazaki fragments* are relatively short (1000 to 2000 bases in prokaryotes) DNA fragments which are synthesized in a discontinuous fashion on the lagging strand during DNA replication. Such fragments appear to be necessary because template DNA is not available for 5' > 3' synthesis until some degree of continuous DNA synthesis occurs on the leading strand in the direction of the replication fork (see K/C-Fig. 10.13). The isolation of such fragments provides support for the scheme of replication shown in K/C-Fig. 10.13.

DNA ligase is required to form phosphodiester linkages in gaps which are generated when DNA polymerase I removes RNA primer and meets newly synthesized DNA ahead of it.

Notice in K/C-Fig. 10.13, the discontinuous DNA strands are ligated together into a single continuous strand.

Primer RNA is formed by RNA primase to serve as an initiation point for the production of DNA strands on a DNA template. None of the DNA polymerases are capable of initiating synthesis without a free 3' hydroxyl group. The primer RNA provides that group and thus can be used by DNA polymerase III.

20. The synthesis of DNA is thought to follow the pattern described in K/C-Fig. 10.13. The model involves opening and stabilization of the DNA helix, priming DNA with synthesis with RNA primer, movement of replication forks in both directions which includes elongation of RNA primers in continuous and discontinuous 5' > 3' modes. Okazaki fragments generated in the replicative process are joined together with DNA ligase. DNA gyrase relieves supercoils generated by DNA unwinding.

21. Eukaryotic DNA is replicated in a manner which is very similar to that of *E. coli*. Synthesis is bidirectional, continuous on one strand and discontinuous on the other, and the requirements of synthesis (four deoxyribonucleoside triphosphates, divalent cation, template, and primer) are the same. Okazaki fragments of eukaryotes are about one-tenth the size of those in bacteria.

Because there is a much greater amount of DNA to be replicated and DNA replication is slower, there are multiple initiation sites for replication in eukaryotes in contrast to the single replication origin in prokaryotes. Replication occurs at different sites during different intervals of the S phase. The proposed functions of four DNA polymerases are described in the text.

22.

(a) Similarity of nearest neighbor frequencies will not yield information as to the amount of DNA in a sample because only the "distribution of neighbors" is being estimated.

(b) In the spleen phosphodiesterase procedure for determining nearest neighbor frequencies, labeled 3'-nucleoside phosphates are generated. The percentage of label (^{32}P) in each nucleoside phosphate is an estimate of the percentage of each class of nucleoside phosphate and as such represents the base composition of each DNA.

(c) Nearest neighbor frequencies do not provide information as to base sequences. Even though the nearest neighbor frequencies between two DNA strands may be *similar*, sequences can vary.

23.

Initial Labeled Base	Labeled Base After Spleen Phosphodiesterase digestion	
	ANTIPARALLEL	PARALLEL
G	C, T, C	C, T,A
C	T,A	T,A,G
T	T,A,G,A	T,A,G
A	C,A,G	C,A,T

Interestingly, one can determine which model occurs in nature by comparing the pattern in which the labeled phosphate is shifted after spleen phosphodiesterase digestion. Focus your attention on the antiparallel model and notice that the frequency which "C" (for example) is the 5' neighbor of "G" is not necessarily the same as the frequency which "G" is the 5' neighbor of "C". However, in the parallel model (b), the frequency which "C" is the 5' neighbor of "G" is the same as the frequency which "G" is the 5' neighbor of "C". By examining such "digestion frequencies" it can be determined that DNA exists in the opposite polarity. See discussion in answer #24.

24. In order to discuss this problem, a symbolism of C->A will signify that C has an A as its 5' neighbor and the labeled phosphate is transferred from C to A by spleen phosphodiesterase digestion.

(a) If there are no polarity restrictions, then the T->T frequency should equal the A->A frequency as long as the DNA is double stranded and the AT and GC base pairings hold. Notice in the figures in Problem #23, the TT/AA base pairs. In digestion of that region with spleen phosphodiesterase, regardless of polarity the T->T frequency should equal the A->A frequency.

(b) By examing the figure in Question #23, one can see that under a *parallel* model, the C->T frequency, for example, should be the same as the G->A frequency, which is not the case. In addition, the C->A frequency should be the same as the G->T fequency, which is not the case. Under an *antiparallel* model the GC and AT base complementation requires that, for example, the C->T frequency should be the same as the A->G frequency. Likewise, the C->A frequency should be the same as the T->G frequency. Following the data in the table, this appears to be the case.

(c)   Already described:

C -> T   should equal A -> G ***
C -> A   should equal   T -> G ***

Additional relationships are as follows:

G ->   A   should equal T -> C
G ->   T   should equal A -> C ***
T ->   G   should equal C -> A ***
G ->   G   should equal C -> C ***
T ->   T   should equal A -> A ***

*** indicates that the expectation is realized. The data do not meet expectation for G -> A as compared to T -> C.

25. Because the two inner strands are essentially exchanging their pairing partners, they must be in the same polarity if they are to maintain the proper antiparallel orientation with the two outer strands.

11

The Genetic Code

Vocabulary: Organization and listing of terms

Structures and Substances

Codon

Messenger RNA

Proflavin

Phage T4

Strain B of *E. coli*

Strain K12 of *E. coli*

Polynucleotide phosphorylase

    random assembly of nucleotides

Homopolymer codes

    RNA homopolymers

    RNA heteropolymers

N-formylmethionine (fmet)

Cell-free protein-synthesizing system

    ribosomes, tRNAs, amino acids, etc.

    artificial mRNAs

Triplet binding assay

Repeating copolymers

Suppression

    change in anticodon

    aminoacyl synthetases

    ribosomal proteins

Overlapping genes

    multiple initiation

Processes/Methods

Transcription, Translation

Frameshifts

Concepts

Genetic code

    triplet codon

        experimental support

frameshift mutations (r_{II})

(+++)(- - -)

codon assignments

   artificial mRNAs

      homopolymers

      mixed copolymers

      % amino acid incorporation

   triplet binding assay

   repeating copolymers

confirmation of codon assignments

   MS2 sequencing

   colinear relationship

unambiguous

degenerate

   support for degenerate code

      "absence of nonsense triplets"

punctuation

   start

      initiator codon

N-formylmethionine (fmet)

AUG, GUG (rare)

stop

   UAA, UAG, UGA

   nonsense mutation

      amber, ochre, opal

      commaless

nonoverlapping

   support for nonoverlapping code

      limitations of tripeptide sequences

      single amino acid substitutions

      adaptor molecule

universal

   yeast, human mitochondrial DNA

Hypotheses

   diamond code

   adaptor molecule

   messenger RNA

   wobble hypothesis

      pattern of degeneracy

Solutions to Problems and Discussion Questions

1. To visualize the limitations which Brenner suggests attend an overlapping code, (only 16 different tripeptides) consider a sequence of five bases as in 1 2 3 4 5. Occupy bases 2, 3, and 4 with the specification of one amino acid. Given that there are four kinds of bases, position 1 can vary four ways. The same may be said for position 5. Thus there will be the possibility of 4 X 4, or 16 variations of the 1 2 3 4 5 sequence with the three middle bases fixed. These 16 variations can, at most, call in 16 different sequences of amino acids.

Since the code is actually nonoverlapping, each position of an amino acid (each triplet) is free to vary, therefore if one considers keeping the central amino acid fixed, there will be 20 X 20, or 400 possible different tripeptides. If the middle amino is free to vary, there are 20 X 20 X 20 possible tripeptides.

2. In order to arrive at the 20 different conformations, place all the combinations of bases in each of the diamonds presented in the problem. The combinations to be placed are as follows: CC, GG, TT, AA, CG, CT, CA, GT, GA, TA.

3. No. The the term "reading frame" refers to the number of bases contained in each codon. The reason that (+++) or (- - -) restored the reading frame *is because* the code is triplet. By having the (+++) or (- - -), the translation system is "out of phase" until the third "+" or "-" is encountered. If the code contained six nucleotides (a sextuplet code), then the translation system is "out of phase" until the sixth "+" or "-" is encountered.

4. If one *only* had data involving 6 (+)'s or 6 (-)'s, then it would be impossible to distinguish between a triplet or sextuplet code. However, if one compared these data to those received from 3(+)'s or 3(-)'s, then it is likely that, given a triplet code, the *frequency* of reading frame restorations which produced a "wild type" phenotype would be higher with the 3(+)'s or 3(-)'s manipulations.

5. **(a)** The way to determine the fraction which each triplet will occur with a random incorporation system is to determine the likelihood that each base will occur in each position of the codon (first, second, third), then multiply the individual probabilities (fractions) for a final probability (fraction).

$$GGG \ = \ 3/4 \ X \ 3/4 \ X \ 3/4 \ = \ 27/64$$

$$GGC \ = \ 3/4 \ X \ 3/4 \ X \ 1/4 \ = \ 9/64$$

$$GCG \ = \ 3/4 \ X \ 1/4 \ X \ 3/4 \ = \ 9/64$$

$$CGG \ = \ 1/4 \ X \ 3/4 \ X \ 3/4 \ = \ 9/64$$

$$CCG \ = \ 1/4 \ X \ 1/4 \ X \ 3/4 \ = \ 3/64$$

$$CGC \ = \ 1/4 \ X \ 3/4 \ X \ 1/4 \ = \ 3/64$$

$$GCC \ = \ 3/4 \ X \ 1/4 \ X \ 1/4 \ = \ 3/64$$

$$CCC \ = \ 1/4 \ X \ 1/4 \ X \ 1/4 \ = \ 1/64$$

(b) Glycine:

GGG and one G_2C (adds up to 36/64)

Alanine:

one G_2C and one C_2G (adds up to 12/64)

Arginine:

one G_2C and one C_2G (adds up to 12/64)

Proline:

one C_2G and CCC (adds up to 4/64)

(c) With the wobble hypothesis, variation can occur in the third position of each codon.

Glycine: GGG, GGC

Alanine: CGG, GCC, CGC, GCG

Arginine: GCG, GCC, CGC, CGG

Proline: CCC, CCG

6. First, compute the frequency (percentages would be easiest to compare) for each of the random codons.

| For 4/5 C: 1/5 A: |

CCC= 4/5 X 4/5 X 4/5   =   64/125 (51.2%)

C_2A = 3(4/5 X 4/5 X 1/5) = 48/125 (38.4%)

CA_2 = 3(4/5 X 1/5 X 1/5) = 12/125 (9.6%)

AAA= 1/5 X 1/5 X 1/5   =   1/125 (0.8%)

| For 4/5 A: 1/5 C: |

AAA= 4/5 X 4/5 X 4/5   =   64/125 (51.2%)

A_2C = 3(4/5 X 4/5 X 1/5) = 48/125 (38.4%)

AC_2 = 3(4/5 X 1/5 X 1/5) = 12/125 (9.6%)

CCC= 1/5 X 1/5 X 1/5   =   1/125 (0.8%)

Proline: C_3, and one of the C_2A triplets

Histidine: one of the C_2A triplets

Threonine: one C_2A triplet, and one A_2C triplet

Glutamine: one of the A_2C triplets

Asparagine: one of the A_2C triplets

Lysine: A_3

7. List the substitutions, then from the code table, apply the codons to the original amino acids. Select codons which provide single base changes.

Original		*Substitutions*
threonine	—>	*alanine*
<u>A</u>C(U,C,A, or G)		<u>G</u>C(U,C,A, or G)
glycine	—>	*serine*
<u>G</u>G(U or C)		<u>A</u>G(U or C)
isoleucine	—>	*valine*
<u>A</u>U(U, C or A)		<u>G</u>U(U, C or A)

8. Assume that you have introduced a copolymer (ACACACAC...) to a cell free protein synthesizing system. There are two possibilities for establishing the reading frames:

ACA, if one starts at the first base and CAC if one starts at the second base. These would code for two different amino acids (ACA = threonine; CAC = histidine) and would produce repeating polypeptides which would alternate *thr-his-thr-his...* or *his-thr-his-thr...*

Because of a triplet code, a trinucleotide sequence will, once initiated, remain in the same reading frame and produce the same code all along the sequence regardless of the initiation site.

Given the sequence CUACUACUACUA, notice the different reading frames producing three different sequences each containing the same amino acid.

Codons:	CUA	CUA	CUA	CUA...
Amino Acids:	leu	leu	leu	leu...
	UAC	UAC	UAC	UAC...
	tyr	tyr	tyr	tyr...
	ACU	ACU	ACU	ACU...
	thr	thr	thr	thr...

If a tetranucleotide is used, such as ACGUACGUACGU...

Codons:	ACG	UAC	GUA	CGU	ACG...
Amino Acids:	thr	tyr	val	arg	thr...
	CGU	ACG	UAC	GUA	CGU...
	arg	thr	tyr	val	arg...
	GUA	CGU	ACG	UAC	GUA..
	val	arg	thr	tyr	val...
	UAC	GUA	CGU	ACG	UAC...
	tyr	val	arg	thr	tyr...

Notice that the sequences are the same except that the starting amino acid changes.

9. The UUACUUACUUAC tetranucleotide sequence will produce the following triplets depending on the initiation point: UUA = leu; UAC = tyr; ACU = thr; CUU = leu. Notice that because of the degenerate code, two codons correspond to leucine.

The UAUCUAUCUAUC tetranucleotide sequence will produce the following triplets depending on the initiation point: UAU = tyr; AUC = ileu; UCU = ser; CUA = leu. Notice that in this case, degeneracy is not revealed and all the codons produce unique amino acids.

10. From the repeating polymer ACACA... one can say that threonine is either CAC or ACA. From the polymer CAACAA... with ACACA..., ACA is the only codon in common. Therefore threonine would have the codon ACA.

11. As in the previous problem the procedure is to find those sequences which are the same for the first two bases but which vary in the third base. Given that AGG = arg, then information from the AG copolymer indicates that AGA also codes for arg and GAG must therefore code for glu.

Coupling this information from that of the AAG copolymer, GAA must also code for glu, and AAG must code for lys.

12. The enzyme generally functions in the degradation of RNA, however in an *in vitro* environment, with high concentrations of the ribonucleoside diphosphates, the direction of the reaction can be forced toward polymerization.

In vivo, the concentration of ribonucleoside diphosphates is low and the degradative process is favored.

13. The basis of the technique is that if a trinucleotide contains bases (a codon) which are complementary to the anticodon of a charged tRNA, a relatively large complex is formed which contains the ribosome, the tRNA, and the trinucleotide. This complex is trapped in the filter whereas the components by themselves are not trapped. If the amino acid on a charged, trapped tRNA is radioactive, then the filter becomes radioactive.

14. A frameshift mutation is caused by the addition (+) or deletion (-) of a base in DNA. Starting with a (+) frameshift mutation, removal of a base after the (+) will establish the proper reading frame and possibly suppress the mutation and provide a "wild type" phenotype. The same can occur with an initial (-) frameshift mutation and subsequent addition of a base.

An amber mutation (UAG) causes premature termination of a nascent polypeptide chain because there are no corresponding tRNAs (and therefore amino acids) to continue in the "decoding" process. If a mutation can occur in the anticodon of a tRNA such that *an* amino acid, perhaps any amino acid, can be placed in the growing polypeptide chain, then the chain can be produced. Suppression occurs when the "repaired" polypeptide chain produces a phenotype which is similar or identical to wild type.

15. Generally, the changes between the "normal" and "altered" code words are due either to changes in the third base recognition or by increased wobble in the codon-anticodon base pairing. The folding of certain mitochondrial tRNAs may be altered to account for differences in base pairing. In some cases a change occurs in the first position; however, in the replacement of *leu* by *thr* in yeast mitochondria, two base changes occur, thus making this change most dramatic and unsusual.

16. Apply the most conservative pathway of change.

17, 18. Because Poly U is complementary to Poly A, double-stranded structures will be formed. In order for an RNA to serve as a messenger RNA it must be single-stranded thereby exposing the bases for interaction with ribosomal subunits and tRNAs.

19. Applying the coding dictionary K/C-Fig. 11.8, the following sequences are "decoded"

Sequence 1:   met-pro-asp-tyr-ser-(term)

Sequence 2:   met-pro-asp-(term)

The 12th base (a uracil) is deleted from Sequence # 1 thereby causing a frameshift mutation.

20. Given the sequence GGA, by changing each of the bases to the remaining three bases, then checking the code table, one can determine whether amino acid substitutions will occur.

G G A > gly	G G U > gly
U G A > **term**	G G C > gly
C G A > **arg**	G G A > gly
A G A > **arg**	G G G > gly
G U A > **val**	U G U > **cys**
G C A > **ala**	C G U > **arg**
G A A > **glu**	A G U > **ser**
G G U > gly	G U U > **val**
G G C > gly	G C U > **ala**
G G A > gly	G A U > **asp**

21.

(a) Starting from the 5' end and locating the AUG triplets one finds two initiation sites leading to the following two sequences:

met-his-thr-tyr-glu-thr-leu-gly

met-arg-pro-leu-gly

(b) In the shorter of the two reading sequences (the one using the internal AUG triplet), a UGA triplet was introduced at the second codon. While not in the reading frames of the longer polypeptide (using the first AUG codon) the UGA triplet eliminates the product starting at the second initiation codon.

22. By examining the coding dictionary K/C-Fig. 11.8 one will notice that the number of codons for each particular amino acid (synonyms) is directly related to the frequency of amino acid incorporation stated in the problem.

12

Synthesis of RNA and Protein

Vocabulary: Organization and listing of Terms

Structures and Substances

Polypeptide

Ribosome

  rRNA

  5S RNA

  5.8S RNA

  subunits

  ribosomal proteins

  nucleolus

  modified bases

  monosome (70S, 80S)

  small subunit

    binding sites

      mRNA, tRNA

      initiation factors

  large subunit

    peptidyltransferase regions

    elongation factors binding sites

      exit point for newly formed chain

Ribosome complex

  peptidyl (P site)

  aminoacyl (A site)

  peptidyl transferase

Adaptor molecules (transfer RNAs - tRNA)

Messenger RNA (mRNA)

RNA polymerase (*E. coli*) -

  holoenzyme (α, β, β', σ)

RNA polymerase (eukaryotic) - I, II, III

  nucleoside triphosphates (NTPs)

  nucleoside monophosphates (NMPs)

  nucleotides

  promoters (promoter sequences)

Consensus sequences

  -10 Pribnow box (TATAAT)

  non-template strand

-35 sequence (TTGACA)

adenine and thymine richness

Goldberg-Hogness (-25, TATA box)

-50 to -500 noncoding regions

    (modulation)

  GC richness

  CCAAT sequence

enhancers

TATA-factor (TFIID)

tRNA (4S)

  codon/anticodon

  precursors

    modified bases

  cloverleaf model

    paired and unpaired loops

    ...pCpCpA (3')

    ...pG (5')

  aminoacyl tRNA synthetases

    charging

    activated form

      (aminoacyladenylic acid)

Initiation factors

  initiation complex

  Shine-Dalgarno sequence

Formylmethionine (N-formylmethionine)

  $tRNA^{fmet}$

Peptidyl transferase

Elongation factors

GTP-dependent release factors

Polyribosomes (polysomes)

Heterogeneous RNA (hnRNA)

  pre-mRNAs

  split genes (intervening sequences)

    introns

    exons

    heteroduplexes

  poly-A

  cap (7mG)

    5' to 5'

Processes/ Method

Transcription

  template binding

  denaturation (unwinding)

  DNA footprinting (F12.1)

  chain elongation (5' to 3')

  chain termination

    termination factor (ρ)

Translation

   tRNA charging

      aminoacyl tRNA synthetases

      charging

      activated form

         (aminoacyladenylic acid)

   chain initiation

   chain elongation

      translocation

   chain termination

      UAG, UAA, UGA

RNA processing

   split genes

   post-transcriptional changes

      poly-A (3')

      cap (5')

   mechanisms

   tRNA molecular splicing

      rRNA self-excision (ribozyme)

      mtDNA (RNA maturase)

      mRNA (spliceosome)

         snRNAs, SNURPS

      lariat

Simultaneous transcription and translation

   (prokaryotes) (F12.2)

"Heavy" isotopes

Molecular hybridization

Sucrose gradients

Absorbance at 260nm

Electron microscopy

   heteroduplexes

Concepts

Information flow (F12.3)

   transcription

      an intermediate molecule

      RNA polymerases (I, II, III)

         sensitivity to α-amanitin

      consensus sequences

         -10 sequence (Pribnow box)

   translation (F12.3)

   gene amplification

      moderately repetitive sequences

      redundancy (tandem repeats)

         spacer DNA

   RNA splicing

      beta-globin gene

      ovalbumin gene

      pro-α-2(I) collagen

      mechanisms (4 classes)

F12.1. The technique of DNA footprinting allows one to identify regions of nucleic acids which interact with proteins. Proteins protect regions from endonucleases.

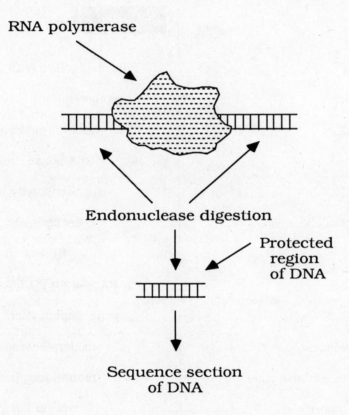

RNA polymerase

Endonuclease digestion

Protected region of DNA

Sequence section of DNA

F12.2 Polarity constraints associated with simultaneous transcription and translation. The RNA polymerase is moving downward (bold arrow) in this drawing.

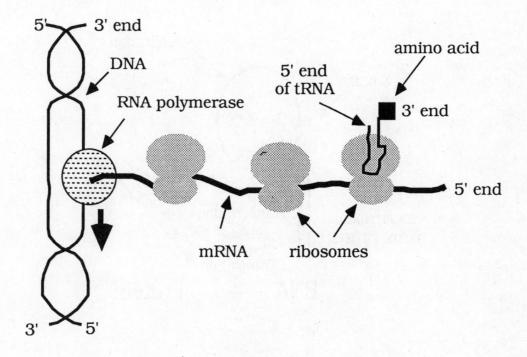

F12.3. Illustration of the processes, transcription and translation, involved in protein synthesis. Such relationships are often called the **Central Dogma**.

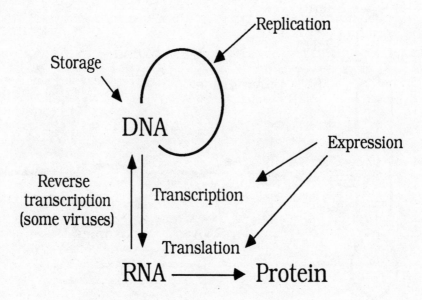

Solutions to Problems and Discussion Questions

1. The central dogma of molecular genetics and to some, all of biology, states that DNA produces, through transcription, RNA, which is "decoded" (during translation) to produce proteins. See F12.3 for a graphic description.

2. Several observations indicated that a "messenger" molecule exists.

First, DNA, the genetic material, is located in the nucleus of a eukaryotic cell whereas protein synthesis occurs in the cytoplasm. DNA therefore does not directly participate in protein synthesis.

Second, RNA, which is chemically similar to DNA is synthesized in the nucleus of eukaryotic cells. Much of the RNA migrates to the cytoplasm, the site of protein synthesis.

Third, there is generally a direct correlation between the amounts of RNA and protein in a cell. More direct support was derived from experiments showing that an RNA other than that found in ribosomes was involved in protein synthesis and shortly after phage infection, an RNA species is produced which is complementary to phage DNA.

3. RNA polymerase from *E. coli* is a complex, large (almost 500,000 daltons) molecule composed of subunits (α, β, β', σ) in the proportion α2, β, β', σ for the holoenzyme. The β subunit provides catalytic function while the sigma (σ) subunit is involved in recognition of specific promoters. The core enzyme is the protein without the sigma subunit.

4. Ribonucleoside triphosphates and a DNA template in the presence of RNA polymerase and a divalent cation (Mg⁺⁺) produce a ribonucleoside monophosphate polymer, DNA, and pyrophosphate (diphosphate). Equimolar amounts of precursor ribonucleoside triphosphates, and product ribonucleoside monophosphates and pyrophosphates (diphosphates) are formed.

5. In *E. coli* transcription and translation can occur simultaneously. Ribosomes add to the 5' end of nascent mRNA and progress to the 3' end during translation. While transcription/translation can be "visualized" in *E. coli* (F12.2), the predominant components "visualized" are the strings of ribosomes (polysomes).

Simultaneous transcription and translation is not visualized in the newt, *Notophthalmus viridescens*, because, being a eukaryote, considerable molecular distance occurs between the sites of transcription (nuclear) and translation (cytoplasmic). What one sees in the newt is a group of rRNA products which are partially complexed with proteins.

6. A functional polyribosome will contain the following components (see K/C-Figs. 12.3, 12.9 and 12.10):

mRNA, charged tRNA, large and small ribosomal subunits (see K/C-Fig. 12.3), elongation and perhaps initiation factors, peptidyl transferase, GTP, Mg⁺⁺, nascent proteins, possibly GTP-dependent release factors.

7. The three main steps in translation are chain initiation, elongation, and termination. *Initiation* involves the formation of the initiation complex as shown in K/C-Fig. 12.9. *Elongation* involves the binding of charged tRNAs, translocation, and the formation of the growing polypeptide chain (K/C-Fig.12.10). *Termination* is presented in K/C-Fig.12.11. It involves release factors, and UAG, UAA, and/or UGA codes.

8. Transfer RNAs are "adaptor" molecules in that they provide a way for amino acids to interact with sequences of bases in nucleic acids. Amino acids are specifically and individually attached to the 3' end of tRNAs which possess a three-base sequence (the anticodon) to base-pair with three bases of mRNA. Messenger RNA, on the other hand, contains a copy of the triplet codes which are stored in DNA. The sequences of bases in mRNA interact, three at a time, with the anticodons of tRNAs.

9. Enzymes involved in transcription include the following: RNA polymerase (*E. coli*), RNA polymerase I, II, III (eukaryotes). Those involved in translation include the following: aminoacyl tRNA synthetases, peptidyl transferase, and GTP-dependent release factors.

10. It was reasoned that there would not be sufficient affinity between amino acids and nucleic acids to account for protein synthesis. For example, acidic amino acids would not be attracted to nucleic acids. With an adaptor molecule, specific hydrogen bonding could occur between nucleic acids, and specific covalent bonding could occur between an amino acid and a nucleic acid tRNA.

11. The sequence of base triplets in mRNA constitutes the sequence of codons. A three-base portion of the tRNA constitutes the anticodon.

12. The small subunit of the ribosome in the presence of initiation factors initially associates with the 5' end of the mRNA. As shown in K/C-Fig. 12.9, the tRNA^fmet then associates at the P site, followed by the addition of the large subunit. Additional charged tRNAs associate with the initiation complex and translocation and elongation occurs.

The large subunit contains regions for binding of the peptidyl transferase and elongation factors. It also serves as the exit point for the newly formed polypeptide chain. At termination, the polypeptide chain is cleaved from the ribosome and the two ribosomal subunits separate.

13. The following is derived from information found in K/C-Fig. 12.3:

	Prokaryotes	Eukaryotes
Monosome	70S	80S
Subunits	50S, 30S	60S, 40S
rRNA		
large	23S + 5S	28S + 5S + 5.8S
small	16S	18S
Proteins		
large	31	~ 50
small	21	~ 30

14. *Fidelity* refers to the accuracy or preciseness of a process. As with DNA replication, errors are made in the production of proteins. For instance, it is estimated that during chain elongation in translation, there is an error rate of 10^{-4} so that approximately one polypeptide in 20 (average length of 500 amino acids) carries one incorrect amino acid. Additionally, because many eukaryotic RNAs are spliced, "splicing errors" are likely.

Because defective proteins are difficult to assay, both qualitatively and quantitatively, it is impossible to determine the exact percentage of incorrectly manufactured proteins. It is likely that "back-up" and repair systems exist to minimize errors in essential gene products. As errors increase in frequency, organismic fitness is likely to decrease.

15. Since there are three nucleotides which code for each amino acid, there would be 423 code letters (nucleotides), 426 including a termination codon. This assumes that other features, such as the polyA tail, the 5'cap, and non-coding leader sequences are omitted. Dividing 20 by 0.34, gives the number of nucleotides (about 59) occupied by a ribosome. Dividing 59 by three gives the approximate number of triplet codes: approximately 20.

16. While some folding (from complementary base pairing) may occur with mRNA molecules, they generally exist as single-stranded structures which are quite labile. Eukaryotic mRNAs are generally processed such that the 5' end is "capped" and the 3' end has a considerable string of adenine bases. It is thought that these features protect the mRNAs from degradation. Such stability of eukaryotic mRNAs probably evolved with the differentiation of nuclear and cytoplasmic functions.

Because prokaryotic cells exist in a more unstable environment (nutritionally and physically, for example) than many cells of multicellular organisms, rapid genetic response to environmental change is likely to be adaptive. To accomplish such rapid responses, a labile gene product (mRNA) is advantageous. A pancreatic cell, which is developmentally stable (differentiated) and existing in a relatively stable environment could produce more insulin on stable mRNAs for a given transcriptional rate.

17. The steps involved in tRNA charging are outlined in K/C-Fig. 12.8. An amino acid in the presence of ATP, Mg++, and a specific aminoacyl synthetase produces an amino acid-AMP enzyme complex (+ PP$_i$). This complex interacts with a specific tRNA to produce the aminoacyl tRNA.

18. The four sites in tRNA which provide for specific recognition are the following: attachment of the specific amino acid, interaction with the aminoacyl tRNA synthetase, interaction with the ribosome, and interaction with the codon (anticodon). See K/C Fig.12.6.

19. One can conclude that the amino acid is not involved recognition of the anticodon.

20.

(a)

(b) TCCGCGGCTGAGATGA (use complementary bases, substituting T for U)

(c) GCU

(d) Assuming that the AGG... is the 5' end of the mRNA, then the sequence would be

arg-arg-arg-leu-tyr

21. Apply complementary bases, substituting U for T:

(a)

Sequence 1: GAAAAAACGGUA

Sequence 2: UGUAGUUAUUGA

Sequence 3: AUGUUCCCAAGA

(b)

Sequence 1: *glu-lys-thr-val*

Sequence 2: *cys-ser-tyr*

Sequence 3: *met-phe-pro-arg*

(c) Because there are no puncuation codons in Sequence 1, it must be in the middle. Sequence 2, having a termination codon (UGA), must be in the terminal portion. Sequence 3, with the AUG starting codon, must be in the initial portion.

22.

Apply complementary bases: GAAAAAACGGTA

23.

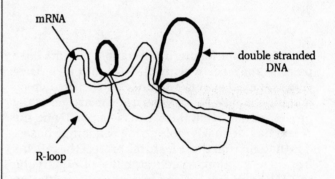

24. Examine K/C-Fig. 12.9 and note that tRNA^fmet enters the P site during the formation of the initiation complex.

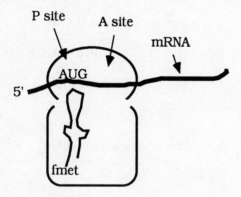

25. The mature β–hemoglobin mRNA will have had the introns removed and when hybridized to the complementary DNA will show a structure similar to that given below:

13

Genes and Proteins

Vocabulary: Organization and Listing of Terms

Structures and Substances

Unit factors

Ferments

Inborn Errors of Metabolism (1909 - Garrod)

    alkaptonuria

        homogentisic acid

    phenylketonuria (PKU)

        phenylalanine hydroxylase

        phenylalanine

        phenylpyruvic acid

    albinism

Arginine biosynthetic pathway

Polypeptide chains

    quaternary structure

    hemoglobin

        heme group

globin portion

    HbA

    HbA_2

    HbF

    Gower 1

    HbC

chains

    chromosome 16

        alpha (α)

        zeta (ζ)

    chromosome 11

        beta (β)

        delta (δ)

        epsilon (ϵ)

        gamma ($^G\gamma$),($^A\gamma$)

    gene families

sickle-cell hemoglobin

    β chain

    6th position (glutamic acid —> valine)

    molecular disease

Polypeptide/Proteins

    carboxyl group (C-terminus)

    amino group (N-terminus)

    R (radical group)

        non-polar (hydrophobic)

        polar (hydrophilic)

        negatively, positively charged

    peptide bond

    dipeptide, tripeptide

    structure

        primary (I°)

        secondary (II°)

        tertiary (III°)

        quaternary (IV°)

            oligomeric

            promoter

    examples

    myoglobin, ribonuclease, fibroin

    hemoglobin, keratin, myosin

    immunoglobins, transport proteins,

hormones, histones, enzymes,

collagen

    tropocollagen

    procollagen

    tripeptide sequences

    prolyl hydroxylase

    lysyl hydroxylase

    scurvy

    ascorbic acid (vitamin C)

    Ehlers-Danlos syndrome

    procollagen peptidase

    osteogenesis imperfecta

        type I

        type II

    Marfan's syndrome

Processes/Method

Metamorphosis

Differentiation

Enzymes

    energy of activation, active site

    catabolic, anabolic, amphibolic

Post-translational modification

Transplantation (imaginal discs)

Electrophoretic migration

    starch gel

    two-dimensional

    anode, cathode

Fingerprinting technique

Recombinant DNA technology

X-ray crystallography

Concepts

One-gene: one-enzyme

One-gene: one-protein

    sickle-cell anemia

    sickle-cell trait

One-gene: one-polypeptide chain

Transplantation (imaginal discs)

    diffusible substances

       cinnabar

       vermilion

    autonomous

Pathway analysis (F13.1)

    Drosophila

       eye color

    Neurospora

       arginine

Gene families

Colinear relationship (nucleotides, proteins)

    synthetase

Structure (F13.2)

    covalent bonds

    hydrogen bonds

       α helix

       β pleated-sheet

    polar, hydrophilic interactions

    nonpolar, hydrophobic interactions

F13.1. An illustration of the relationship between a metabolic block (caused by a mutant gene) and a metabolic pathway. Notice that one mutant gene causes the pathway to be blocked in a specific place.

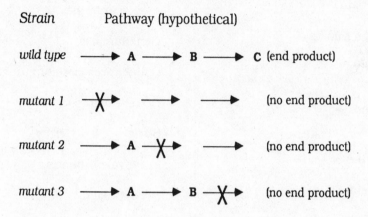

F13.2. Below is a simple sketch of several folding aspects of a protein.

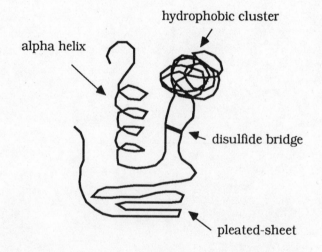

Solutions to Problems and Discussion Questions

1. Phenylalanine is an amino acid which, like other amino acids, is required for protein synthesis. While too much phenylalanine and its derivatives cause PKU, too little will restrict protein synthesis.

2. Both phenylalanine and tyrosine can be obtained from the diet. Even though individuals with PKU can not convert phenylalanine to tyrosine, it is obtained from the diet.

3. Tyrosine is a precursor to melanin, skin pigment. Individuals with PKU fail to convert phenylalanine to tyrosine and even though tyrosine is obtained from the diet, at the population level, individuals with PKU have a tendency for less skin pigmentation.

4. Recall that the term *autonomous* means that the eye color is not influenced by surrounding tissue. *Nonautonomous* means that surrounding tissue (manipulated by transplantation) will influence eye color. Realize that each mutant block influences only one step in a metabolic pathway and that if the host is building up or contains a substance *after* a metabolic block, then the metabolic pathway can be "repaired." For example, because *donor a* is transplanted into *host c* and remains mutant, it is developing autonomously. *Host* c is not able to supply substances after the metabolic block in *donor a*, therefore, the block in c must be before the block in *a*.

The same conclusion may be drawn regarding the relationship between the block produced by gene *a* and the block of gene *b*. From the first two experiments, we can say that blocks *b* and *c* come before block *a*. Likewise, because insertion of *donor c* into *host b* fails to bring about repair, block *b* must be before block *c*. From the first three experiments we can conclude that the blocks are ordered: *b, c, a.*

This tentative answer can be verified by the last three experiments which show that any time a donor imaginal disk is placed in a host where the block is *later* in the pathway, "repair" of the phenotype occurs. The answer therefore is as follows:

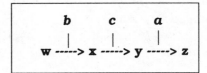

5. Set the gene symbols according to the following convention:

> v = vermilion; v^+ = wild type allele
>
> cn = cinnabar; cn^+ = wild type allele

Label any fly which is

$$v/v, \; v/Y, \; or \; cn/cn$$

as bright red in phenotype.

Females: $v/v; \; cn^+/cn^+$

$$X \qquad v^+/Y; \; cn/cn \; \text{(males)}$$

F_1:

$v^+/v; \; cn^+/cn$ (1/2 wild type females)

$v/Y; \; cn^+/cn$ (1/2 bright red males)

F_2: Using the forked-line method

For the *vermilion* locus For the *cinnabar* locus

1/4 female wild   3/4 wild

1/4 female vermilion 1/4 cinnabar

1/4 male wild

1/4 male vermilion

Multiplying across:

3/16 female wild type

3/16 female bright red

3/16 male wild type

3/16 male bright red

1/16 female bright red

1/16 female bright red

1/16 male bright red

1/16 male bright red

Now collect similar phenotypes:

3/16 female wild type

5/16 female bright red

3/16 male wild type

5/16 male bright red

For the reciprocal cross:

Females: v^+/v^+; cn/cn

      X v/Y; cn^+/cn^+ (males)

F_1:

 v^+/v; cn^+/cn (1/2 wild type females)

 v^+/Y; cn^+/cn (1/2 wild type males)

F_2: Using the forked-line method

For *vermilion* locus For *cinnabar* locus

2/4 female wild   3/4 wild

1/4 male wild   1/4 cinnabar

1/4 male vermilion

Multiplying across:

6/16 female wild type

3/16 male wild type

3/16 male bright red

2/16 female bright red

1/16 male bright red

1/16 male bright red

Now collect similar phenotypes:

6/16 female wild type

2/16 female bright red

3/16 male wild type

5/16 male bright red

6. (a) In this cross, two gene pairs are operating because the F_2 ratio is a modification of a 9:3:3:1 ratio, which is typical of a dihybrid cross. If one assumes that homozygosity for either or both of the two loci gives white, then let strain A be, *aaBB* and strain B, *AAbb*. The F_1 is *AaBb* and pigmented (purple). The typical F_2 ratio would be as follows:

9/16   *A_B_*   Purple

3/16   *aaB_*   white

3/16   *A_bb*   white

1/16   *aabb*   white

If a pathway exists which has the following structure, then the genetic and biochemical data are explained.

```
      aa          bb

   X --------> Y ---------> Purple
(white)      (white)         pigment
```

(b) For this condition, with the pink phenotype present, leave the symbols the same however change the Y compound such that when accumulated, a pink phenotype is produced:

```
      aa          bb

   X --------> Y ---------> Purple
(white)      (pink)          pigment
```

9/16   *A_B_*   Purple

3/16   *aaB_*   white

3/16   *A_bb*   **pink**

1/16   *aabb*   white

7. The best way to approach these types of problems, especially when the data are organized in the form given, is to realize that the substance (supplement) which "repairs" a strain, as indicated by a (+), is *after* the metabolic block for that strain. In addition, and most importantly, is that the substance which "repairs" the highest number of strains either *is the end product* or is *closest to* the end product.

Looking at the table, notice that the supplement tryptophan "repairs" all the strains. Therefore it must be at the end of the pathway or at least after all the metabolic blocks (defined by each mutation). Indole "repairs" the next highest number of strains (3) therefore it must be second from the end. Indole glycerol phosphate "repairs" two of the four strains so it is third from the end. Anthranilic acid "repairs" the least number of strains, so it must be early (first) in the pathway.

Minimal medium is void of supplements and mutant strains involving this pathway would not be expected to grow (or be "repaired"). The pathway therefore would be as follows:

```
AA----->IGP----->I----->TRY
```

To assign the various mutations to the pathway, keep in mind that if a supplement "repairs" a given mutant, the supplement must be after the metabolic block. Applying this rationale to the above pathway, the metabolic blocks are created at the following locations.

```
trp-8     trp-2     trp-3     trp-1
X--|->AA---|->IGP--|->I---|->TRY
```

8. In general the rationale for working with a branched chain pathway is similar to that stated in the previous problem. Since thiamine "repairs" each of the mutant strains, it must, as stated in the problem, be the final synthetic product.

Remembering the "one-gene: one-enzyme" statement, each metabolic block should only occur in one place, so even though pyrimidine and thiazole supplements each "repair" only one strain each, they will not occupy the same step; rather a branched pathway is suggested. Consider that pyrimidine and thiazole are products of distinct pathways an that both are needed to produce the end product, thiamine as indicated below:

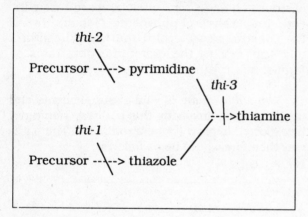

9. The fact that enzymes are a subclass of the general term *protein*, a *one-gene:one-protein* statement might seem to be more appropriate. However, some proteins are made up of subunits, each different type of subunit (polypeptide chain) being under the control of a different gene. Under this circumstance, the *one-gene:one-polypeptide* might be more reasonable.

It turns out that many functions of cells and organisms are controlled by stretches of DNA which either produce no protein product (operator and promoter regions, for example) or more than one function as in the case of overlapping genes and differential mRNA splicing. A simple statement regarding the relationship of a stretch of DNA to its physical product is difficult to formulate.

10. The electrophoretic mobility of a protein is based on a variety of factors, primarily the net charge of the protein and to some extent, the conformation in the electrophoretic environment. Both are based on the type and sequence (primary structure) of the component amino acids of a protein.

The interactions (hydrogen bonds) of the components of the peptide bonds, hydrophobic, hydrophilic, and covalent interations (as well as others) are all dependent on the original sequence of amino acids and take part in determining the final conformation of a protein. A change in the electrophoretic mobility of a protein would therefore indicate that the amino acid sequence had been changed.

11. The following types of normal hemoglobin are presented in the text:

Hemoglobin	Polypeptide chains	
HbA	$2\alpha2\beta$	(alpha, beta)
HbA$_2$	$2\alpha2\delta$	(alpha, delta)
HbF	$2\alpha2\gamma$	(alpha, gamma)
Gower 1	$2\zeta2\epsilon$	(zeta, epsilon)

The alpha and beta chains contain 141 and 146 amino acids, respectively. The zeta chain is similar to the alpha chain while the other chains are like the beta chain.

12. Sickle-cell anemia is coined a *molecular* disease because it is well understood at the molecular level - at the level of a base change in DNA which leads to an amino acid change in the β chain of hemoglobin. It is a *genetic* disease in that it is inherited from one generation to the next. It is not contagious as might be the case of a disease caused by a microorganism. Diseases caused by microorganisms may not necessarily follow family blood lines whereas genetic diseases do.

13. In the late 1940's Pauling demonstrated a difference in the electrophoretic mobility of HbA and HbS (sickle-cell hemoglobin) and concluded that the difference had a chemical basis. Ingram determined that the chemical change occurs in the primary structure of the globin portion of the molecule using the fingerprinting technique. He found a change in the 6th amino acid in the β chain.

14. It is possible for an amino acid to change without changing the electrophoretic mobility of a protein under standard conditions. If the amino acid is substituted with an amino acid of like charge and similar structure there is a chance that factors which influence electrophoretic mobility (primarily net charge) will not be altered. Other techniques such as chromatography of digested peptides may detect subtle amino acid differences.

15. Gene duplication provides an opportunity for genes to change without significantly lowering the fitness of an organism. Under this condition, variation in genes and gene function can accumulate as long as the original and necessary function of the gene is not compromised.

The homology between the β and γ chains indicates that gene duplication and subsequent evolution occurred. Lack of sequence homology between the α and β chains suggests independent origin. These conclusions are supported by the discovery that the β and γ chains are located on the same chromosome (#11) while the α chain is located on chromosome #16.

16. *Colinearity* refers to the sequential arrangement of subunits, amino acids and nitrogenous bases in proteins and DNA, respectively. Sequencing of genes and products in MS2 phage and studies on mutations in the *A* subunit of the *tryptophan synthetase* gene indicate a colinear relationship.

17. "Fine-mapping" meaning precise mapping of mutations *within* a gene, is possible in some phage systems because many recombinants can often be generated relatively easily. Having the precise intragenic location of mutations as well as the ability to isolate the products, especially mutant products, allows scientists to compare the locations of lesions within genes.

Mutations occurring early in a gene will produce proteins with defects near the N-terminus. In this problem, the lesions cause chain termination, therefore the nearer the mutations are to the 5' end of the mRNA, the shorter will be the polypeptide product.

18. As stated in the text, the four levels of protein structure are the following:

Primary: the sequence of amino acids. This sequence determines the higher level structures.

Secondary: α–helix and β-pleated-sheet structures generated by hydrogen bonds between components of the peptide bond.

Tertiary: folding which occurs as a result of interactions of the amino acid side chains. These interactions include, but are not limited to the following: covalent disulfide bonds between cysteine residues; interactions of hydrophilic side chains with water; interactions of hydrophobic side chains with each other.

Quaternary: the association of two or more polypeptide chains. Called *oligomeric*, such a protein is made up of more than one *protomer*.

19. There are probably as many different types of proteins as there are different types of structures and functions in living systems. Your text lists the following:

Oxygen transport: hemoglobin, myoglobin

Structural: collagen, keratin, histones

Contractile: actin, myosin

Immune system: immunoglobins

Cross-membrane transport: a variety of proteins in and around membranes, such as receptor proteins.

Regulatory: hormones, perhaps histones

Catalytic: enzymes

20. Enzymes function to regulate catabolic and anabolic activities of cells. They influence (lower) the *energy of activation* thus allowing chemical reactions to occur under conditions which are compatible with living systems. Enzymes possess *active sites* and/or other domains which are sensitive to the environment.

The active site is considered to be a crevice, or pit, which binds reactants, thus enhancing their interaction. The other domains mentioned above may influence the conformation and therefore function of the active site.

21.

(a) Collagen has a relatively high proportion of glycine and proline residues. The small size of the R-group on glycine facilitates the formation of the triple helical structure.

(b) Relatively long polypeptides, called procollagen, are synthesized by fibroblasts and secreted into extracellular spaces. Procollagen peptidases shorten the procollagen fibers.

(c) The free terminal amino groups of two lysine residues of tropocollagen may react to form a covalent linkage (an aldol cross-link). Other aldol cross-links may form (to histidine residues) to yield a highly stable structure.

(d) The three interacting chains of the triple helix are stabilized by hydrogen bonds. See K/C-Fig. 13.17.

22. Yanofsky's work on the *tryA* locus in bacteria involved the mapping of mutations and the finding that a relationship exists between the position of the mutation in a gene and the amino acid change in a protein. Work with the MS2 by Fiers showed, by sequencing of the coat protein (129 amino acids) and the gene (387 nucleotides), a linear relationship as predicted by the code word dictionary.

23. At position 16 in the alpha chain a change occurs from lysine to aspartic acid. This substitution involves two changes, from AA(A,G) to GA(U,C). All of the other substitutions involve one base change.

14

Mutation and Mutagenesis

Vocabulary: Organization and Listing of Terms

Structures and Substances

Somatic cells

Gamete forming cells

    germ line

    gametes

5-Bromouracil

2-Amino purine

Nitrous acid

Hydroxylamine

Acridine orange

Proflavin

Mustard gas

Ethylmethane sulfonate

ABO antigens

    H substance

        glycosyltransferase

        N-acetyl-α-D-galactosamine

        α-D-galactose

Dystrophin

Pyrimidine dimers

uvr gene product

DNA polymerase I

DNA ligase

Carcinogen (aflatoxin B)

AP endonuclease

DNA glycosylases

GATC sequence

mutH, L, S and *U*

recA

lexA

Fibroblasts

Photoreactivation enzyme

Heterokaryon

Processes/Methods

Variation by mutation

    spontaneous

        background radiation

            cosmic sources

            mineral sources

            ultraviolet light

        rates

    induced

Molecular basis

    base substitution or point mutations

        transition

        transversion

        frameshift

    tautomeric shifts (forms)

    base analogues

        5-bromouracil

        2-amino purine

    reverse mutation

    oxidative deamination

        nitrous acid

    hydroxylation

        hydroxylamine

    alkylation

        mustard gases

        ethylmethane sulfonate

    frameshift mutations

        acridine dyes (acridines)

            acridine orange

        proflavin

Detection

    bacteria and fungi

        minimal medium

        complete medium

        prototrophs

        auxotrophs

        segregation (1:1)

    Drosophila

        ClB

        attached-X

    plants

        visual observation

        tissue culture

    humans

        pedigree analysis

        cell culture (*in vitro*)

        molecular basis

rapid sequencing of DNA

ABO antigens

H substance modification

muscular dystrophy

myopathy

Duchenne muscular dystrophy

Becker muscular dystrophy

dystrophin

Ames test

Salmonella typhimurium

Repair

high-energy radiation

X-rays

gamma radiation

cosmic radiation

free radicals

reactive ions

intensity of dose

roentgen

doubling dose of radiation

ultraviolet radiation (260nm)

pyrimidine dimers

T-T, C-C, T-C

photoreactivation

photoreactivation enzyme (PRE)

excision repair

uvr gene product

DNA polymerase I

DNA ligase

apurinic site

AP endonuclease

DNA glycosylases

proofreading and mismatch repair

strand discrimination

DNA methylation

GATC sequence

mutH, L, S and *U*

recombinational repair

recA

rescue operation

lexA

SOS response

error-prone system

xeroderma pigmentosum (XP)

unscheduled DNA synthesis

DeSanctis-Cacchione syndrome

photoreactivation enzyme

heterokaryon

somatic cell genetics

complementation

Site-directed mutagenesis

Concepts

Mutation - basis of organismic diversity (F14.1)

chromosomal aberrations

gene mutations

somatic (F14.2)

germ line (F14.2)

dominant autosomal

sex-linked recessive

autosomal recessive

morphological

nutritional or biochemical

behavioral

regulatory

lethal

conditional

temperature-sensitive

Toleration of single nucleotide substitutions

Strand discrimination

Complementation

Target theory

Genetic load

F14.1. Graphic representation of the relationship of mutation to Darwinian evolutionary theory.   Mutation provides the original source of variation on which natural selection operates.

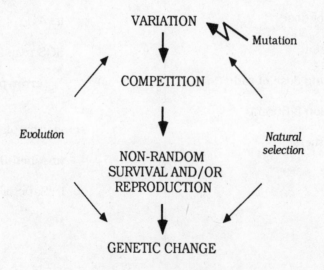

F14.2. Illustration of the difference between somatic and germ-line mutation. Somatic mutations are not passed to the next generation whereas those in the germ line may be passed to offspring.

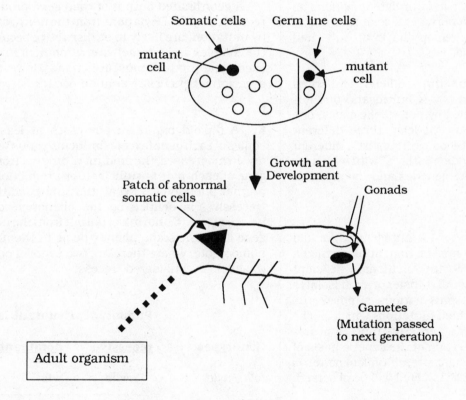

Solutions to Problems and Discussion Questions

1. The term *chromosomal aberration* refers to changes in chromosome number or structure, such as duplications, deletions, inversions, and translocations. A *point mutation* is a change in the nucleotide sequence of a single gene.

2. Mutations are the "windows" through which geneticists look at the normal function of genes, cells, and organisms. When a mutation occurs it allows the investigator to formulate questions as to the function of the normal allele of that mutation. For example, hemophilia is an inherited blood-clotting disease.

Because there are three different inherited forms of the disease, two X-linked and one autosomal, all determined by non-allelic genes, one can say that there are at least three different proteins involved in blood clotting. At a different level, mutations provide "markers" with which biologists can study the genetics and dynamics of populations.

3. As outlined in Chapter 13, mutagenized conidia can be cultured on complete, then minimal media to isolate strains which are nutritionally deficient (auxotrophs). Individual transfer of such isolates to minimal medium plus various supplements allows one to isolate those mutations which can be "repaired." By crossing an auxotroph to a prototroph one can be certain of the genetic basis of the nutritionally deficient strain. Typical patterns of segregation (4:4, 2:2:2:2, etc.) will be observed.

4. It is true that *most* mutations are thought to be deleterious to an organism. A gene is a product of perhaps a billion or so years of evolution and it is only natural to suspect that random changes will probably yield negative results.

However, *all* mutations may not be deleterious. Those few, rare variations which are beneficial will provide a basis for possible differential propagation of the variation. Such changes in gene frequency represent the basis of the evolutionary process. See F14.1 in this book.

5. As stated in the previous problem, a functional sequence of nucleotides, a gene, is likely to be the product of perhaps a billion or so years of evolution. Each gene and its product functions in an environment which has also evolved, or co-evolved.

A coordinated output of each gene product is required for life. Deviations from the norm, caused by mutation, are likely to be disruptive because of the complex and interactive environment in which each gene product must function. However, on occasion a beneficial variation occurs.

6. A diploid organism possesses at least two copies of each gene (except for "hemizygous" genes) and in most cases, the amount of product from one gene of each pair is sufficient for production of a normal phenotype. Recall that the condition of "recessive" is defined by the phenotype of the heterozygote. If one unit of output from the normal gene gives the same phenotype as in the normal homozygote, where there are two units of output, the allele is considered "recessive."

	Phenotype, if mutant is:	
Genotypes	**recessive**	**dominant**
wild/wild	wild	wild
wild/mutant	wild	mutant
mutant/mutant	mutant	mutant

7. A *conditional* mutation is one that produces a wild type phenotype under one environmental condition and a mutant phenotype under a different condition. A conditional *lethal* is a gene which under one environmental condition leads to premature death of the organism.

Conditional
(temperature sensitive)
Mutation

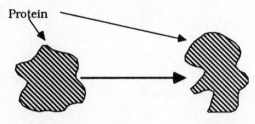

Protein

Normal function at 37 degrees centigrade

Mutant at 42 degrees centigrade

8. Mutations constantly occur in both somatic and gametic tissues. When in somatic tissue, the mutation will not be passed to the next generation, however the physiological or structural role of the mutant cell may be compromised, but of little or no impact. If a somatic mutation occurs early in development and if numerous progeny cells are produced from the mutant cell, then a significant tissue mass may be have abnormal function. In addition, mutation can alter the normal regulatory aspects of a cell cycle. When such regulation is compromised, then abnormal rates of cell proliferation, cancer, may result. Mutations in the gametic tissue may pose problems for future generations. See F14.2 in this book.

9. (a) The primary focus of the *ClB* technique is the isolation of X-linked recessive mutations. One can also determine the frequency of dominant mutations by examining the frequency of F$_1$ *Bar-eyed* females.

(b) The recessive lethal gene eliminates the males which do not have a new recessive lethal mutation. This facilitates scoring of the newly introduced recessive lethals.

(c) Without a crossover suppressor, the newly introduced lethal genes would crossover with the normal chromosome in the F$_1$ *Bar-eyed* females and male F$_2$ offspring would be produced even though a lethal gene had been introduced. Likewise, the benefits of having the recessive lethal gene and the *Bar eye* dominant marker would be lost.

With the crossover suppressor (inversion), a stock can be maintained which would allow additional study of the newly produced mutations.

(d) *C* is an inversion which suppresses the recovery of crossover chromatids.

10. The general expression for computing mutation rate is to divide the number of mutant gametes by the total number of gametes. Because 500 F$_2$ cultures were scored, 500 X chromosomes, carried in 500 X-bearing sperm cells were tested. Since there are many essential genes on the X chromosome and none of them were mutated, one can conclude that the rate of induced mutation for each essential X-linked gene is less than 1/500.

11. Let *II* indicate a mutagenized second chromosome

Females *Cy L/Pm*   X   males *II/II*

F$_1$ males   *Cy L/II*   X   *Cy L/Pm*

(individual crosses)

F$_2$ females   *Cy L/II*   X   *Cy L/II*

(individual crosses)

F_3 genotypes:

Cy L/Cy L   (lethal)

Cy L/ll   (Curly, Lobe)

ll/ll   (dies if recessive lethal)

The *ll/ll* class will be present in crosses where no recessive lethal was introduced. If a recessive lethal had been introduced, only Curly/Lobe flies would be seen in the cultures. In addition, recessive morphological mutations will be expressed in the *ll/ll* offspring.

12. Watson and Crick recognized that various tautomeric forms, caused by single proton shifts, could exist for the nitrogenous bases of DNA. Such shifts could result in mutations by allowing hydrogen bonding of normally noncomplementary bases. As stated in the text, important tautomers involve keto-enol pairs for thymine and guanine, and amino-imino pairs for cytosine and adenine. Such pairings are illustrated in K/C-Fig. 14.6.

13. All four of the agents are mutagenic because they cause base substitutions. Nitrous acid oxidatively deaminates bases such that cytosine is converted to uracil and adenine is converted to hypoxanthine. Uracil pairs with adenine and hypoxanthine pairs with cytosine.

Hydroxylamine adds a hydroxyl group to cytosine, thereby producing hydroxylaminocytosine which may undergo a tautomeric shift and pair with adenine. Alkylating agents donate an alkyl group to the amino or keto groups of nucleotides, thus altering base-pairing affinities. 6-ethyl guanine acts like adenine, thus pairing with thymine. Base analogues such as 5-bromouracil and 2-amino purine are incorporated as thymine and adenine respectively yet they base pair with guanine and cytosine respectively.

14. Frameshift mutations are likely to change more than one amino acid in a protein product because as the reading frame is shifted, new codons are generated. In addition, there is the possibility that a nonsense triplet could be introduced, thus causing premature chain termination. If a single pyrimidine or purine has been substituted, then only one amino acid is influenced.

15. X-rays are of higher energy and shorter wavelength than UV light. They have greater penetrating ability and can create more disruption of DNA.

16. In contrast to UV light, X-rays penetrate surface layers of cells and thus can affect gamete-forming tissues in multicellular organisms. In addition, X-rays break chromosomes and a variety of chromosomal aberrations can result. Ions and free radicals are formed in the paths of X-rays and these interact with components of DNA to cause mutations. UV light generates pyrimidine dimers, primarily thymine, which distort the normal conformation of DNA and inhibit normal function.

17. *Photoreactivation* can lead to repair of UV-induced damage. An enzyme, photoreactivation enzyme, will absorb a photon of light to cleave thymine dimers. *Excision repair* involves the products of several genes, DNA polymerase I, and DNA ligase to clip out the UV-induced dimer, fill in, and join the phosphodiester backbone in the resulting gap. The excision repair process can be activated by damage which distorts the DNA helix.

Recombinational repair is a system which responds to DNA which has escaped other repair mechanisms at the time of replication. If a gap is created on one of the newly synthesized strands a "rescue operation or SOS response" allows the gap to be filled. Many different gene products are involved in this repair process: *recA, lexA*. In SOS repair, the normal proofreading of DNA polymerase III is suppressed and this therefore is called an "error-prone system."

18. Because mammography involves the use of X-rays and X-rays are known to be mutagenic, it has been suggested that frequent mammograms may do harm.

19. Because recessive genes are "hidden" in the heterozygote, their frequency is impossible to determine immediately. Such recessive genes may reveal themselves in the homozygous state (as visibles or lethals) in future generations and as such may constitute a considerable genetic and obviously emotional load.

20. In the *Ames assay* the compound to be tested is incubated with a mammalian liver extract to simulate an *in vivo* environment. This solution is then placed on culture plates with an indicator microorganism, *Salmonella typhimurium*, which is defective in its normal repair processes. The frequency of mutations in the tester strains is an indication of the mutagenicity of the compound.

21. In *site-directed mutagenesis* the goal is usually to change one or more of the nucleotides within a gene. A piece of DNA complementary to the gene being studied is obtained which contains the desired altered base sequence. The DNA with the altered base sequence will hybridize with the original (unaltered) DNA strand and upon replication, two types of duplexes will be formed, one like the original, unaltered sequence, the other will contain the altered sequence. The altered gene sequence can be used to transform competent bacteria.

22. *Xeroderma pigmentosum* is a form of human skin cancer caused by perhaps several rare autosomal genes which interfere with the repair of damaged DNA. Studies with heterokaryons provided evidence for complementation, indicating that there may be as many as seven different genes involved.

23. Much of what is known about UV-induced mutation repair comes from studies on human fibroblasts in culture. Such cells show unscheduled DNA synthesis typical of DNA repair, excision repair, and photoreactivation. Individuals with xeroderma pigmentosum are unable to utilize one or more of these repair mechanisms.

24. In *excision repair* a small section of DNA is removed, and subsequently "filled-in" by DNA polymerase activity. Such represents "unscheduled DNA synthesis." One can determine complementation groupings by placing each heterokaryon giving a "0" into one group and those giving a "+" into a separate group. For instance, XP1 and XP2 are placed into the same group because they do not complement each other. However, XP1 and XP5 do complement ("+") therefore they are in a different group. Completing such pairings allows one to determine the following groupings:

XP1	*XP4*	*XP5*
XP2		*XP6*
XP3		*XP7*

The groupings (complementation groups) indicate that there are at least three "genes" which form products necessary for unscheduled DNA synthesis. All of the cell lines which are in the same complementation group are defective in the same product.

25. Each individual arises from the union of two gametes. If there are 100,000 genes per genome (haploid) then the number of new mutations per individual would be as follows:

$$2(5 \times 10^{-5})(1 \times 10^5) = 10$$

Assuming 4.3×10^9 individuals, there would be 4.3×10^{10} new mutations in the current populace.

26. Given that the cells were treated, then allowed to complete one round of replication, the final computation of the mutation rate should be divided by two. The general expression for the mutation rate is the number of mutant cells divided by the total number of cells. In this case the equation would be as follows:

$$\frac{18 \times 10^1}{6 \times 10^7}$$

or 3×10^{-6}

Now dividing by two (as stated above) gives

1.5×10^{-6}.

Part Two: Sample Test Questions (with detailed explanations of answers)

Question 1. The foundations of molecular genetics rest upon the assumption that a genetic material exists with the following properties:

a. Autocatalytic (can replicate itself)

b. Heterocatalytic (can direct form and function)

c. Mutable

d. Can exist in an infinite number of forms

(a) Provide a simple sketch which demonstrates the replication scheme of DNA.

(b) Briefly describe how DNA provides form and function.

(c) At the level of nucleotides, what characterizes mutant DNA?

(d) Why may we say that DNA can exist in an infinite number of forms?

Concepts:

understanding of gene function

understanding DNA structure

function

mutation

variation

Answer 1.

(a) DNA replicates in a semiconservative manner such that each daughter strand is "half-new" and "half-old" in a particular pattern.

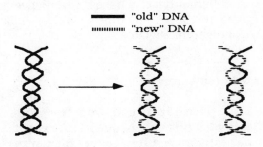

(b) The *Central Dogma of Biology* is based on the production (through transcription) of a RNA messenger from a DNA template and the subsequent "decoding" of that messenger by the process of translation.

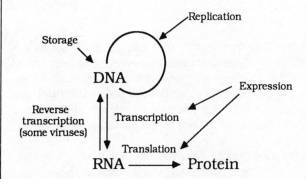

(c) The DNA template contains a sequence of nitrogenous bases which specifies a code from which amino acids are ordered in proteins. Through tautomeric shifts and a number of other natural factors (radiation, chemicals) changes can occur in that sequence of bases. Indeed, the mechanisms by which genes replicate themselves generates errors and leaves us with the conclusion that DNA is an inherently unstable molecule.

(d) Given the variety of organisms and the variation within organisms, there must be numerous, hundreds of millions, elementary factors which are inherited. DNA can provide for this variety by differences in the length and sequence of bases for each inherited functional unit. Given that there are four different types of bases, a sequence having merely ten bases would be capable of 4^{10} (over 1 million) different sequences.

Common errors:

There are usually very few problems with this type of question, except that students often have difficulty clearly explaining that which they know in model form. Written descriptions tend to be more lists of examples rather than explanations of structures and/or processes.

Question 2. Assume that you are microscopically examining mitotic metaphase cells of an organism with a 2N chromosome number of 2 (both telocentric). Assume also that the cell passed through one S phase labeling (innermost phosphate of dCTP radioactive) just prior to the period of observation.

(a) Draw this cell's chromosomes, and the autoradiographic pattern you would expect to see.

(b)Assuming that the A+T/G+C ratio of the DNA in this cell is 1.67 and this DNA is digested with snake venom diesterase (cleaves at the 3' position), what percentages of the total radioactivity would the following products have?

     Adenine_____   Guanine_____

     Thymine_____   Cytosine_____

Concepts:

    semiconservative replication

    chromosome morphology

    DNA structure

       labeling

       enzymatic digestion

       5', 3' orientations

Answer 2.

(a) There will be two telocentric metaphase chromosomes in the drawing and each chromatid will be labeled.

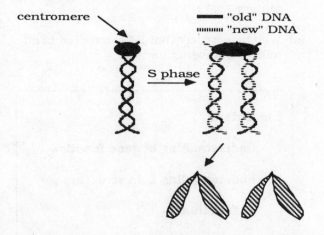

This autoradiographic pattern results because of semiconservative replication. Any cell which contains labeled chromosomes will have had their chromosomes pass through an S phase in the presence of label.

(b) Consider that the DNA was labeled with a dCTP having the innermost phosphate labeled. As this triphosphonucleotide is incorporated into the DNA, it will have the following relationship to its neighbors. *Snake venom diesterase* cleaves DNA at the 3' position meaning that it breaks the bond between the phosphate and the 3' carbon.

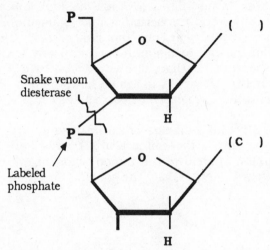

Therefore, the labeled phosphate remains attached to the 5' carbon of the cytosine nucleotide. The A+T/G+C ratio of 1.67 is of no consequence in answering this problem because all of the label remains attached to the cytosine.

Adenine_____   Guanine_____

Thymine_____   Cytosine _100%_

Common errors:

inappropriate labeling pattern

inability to draw telocentric chromosomes

understanding cleavage at 3' position

eliminating extraneous information

Question 3. Below is a schematic of transcription and translation occurring simultaneously as described by Miller *et al.* (1970) in *E. coli.* The broken circle symbolizes RNA polymerase. Answer the questions below,

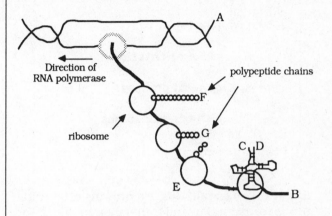

1. Is "A" at the 5' or 3' end of the DNA strand?
_____

2. Is "B" at the 5' or 3' end of the RNA strand?
_____

3. Is "C" at the 5' or 3' end of the RNA strand?
_____

4. What type of RNA is closest to letter "D"?
_____

5. Would base sequences near letters "C" or "D" (state which) be expected to hold the amino acid?
_____

6. What is the S value of the rRNA in the small subunit of the ribosome closest to letter "E"?_____

7. Is the amino acid nearest letter "F" the same type as the one nearest letter "G" (yes or no)?
_____

Concepts:

> **transcription**
>
> > **5', 3' orientations**
>
> **translation**
>
> > **tRNA orientation**
> >
> > **general process**
> >
> > **rRNA in ribosomes**

Answer 3. Overall, this is a drawing of simultaneous transcription and translation in which the RNA polymerase is moving from right to left, making a mRNA which is complementary to one of the two strands of DNA. Ribosomes have added to the nascent (newly forming) mRNA and what appears to be polypeptide chains are protruding from the ribosomes.

(1, 2) In answering this question remember that all synthesis of nucleic acids starts at the 5' end and finishes at the 3' end. Therefore, immediately label the end near point "B" with a 5'. The is the nascent mRNA. Recall that all orientation of complementary strands is antiparallel and since the RNA polymerase is going from right-to-left (according to the arrow) the end of the DNA strand from which the mRNA is copied is the 3' end. Now, since the 3' end of the DNA template strand is identified, its DNA complementary end (at point "A") must be the 5' end.

(3) The codon-anticodon relationship is also antiparallel (based on hydrogen bonding) and since the 5' end of the mRNA is identified, letter "C" must be at the 5' end.

(4) While the diagram is *not to scale*, given the folded structure of the molecule and its position in the ribosome, consider the RNA nearest to letter "D" as tRNA.

(5) Remember that the 3' end of the tRNA holds the amino acid, therefore letter "D" is where the amino acid would be attached.

(6) The tRNA binds mainly to the large subunit of the ribosome, while the mRNA binds mainly to the small subunit of the ribosome. The question asks for the S value of the rRNA in that small subunit. Simultaneous transcription and translation occurs in prokaryotes only (not eukaryotes). The S value for the small subunit of a prokaryotic ribosome is 30S but that value includes both rRNA and protein. The rRNA molecule however has an S value of 16, which is the correct answer.

(7) Since all of the ribosomes are moving along the same mRNA, the amino acid sequences are the same. Therefore, the amino acids neares the letters "F" and "G" are the same.

Common errors:

> **polarity of DNA and RNA strands**
>
> **structure of ribosomes**
>
> **overall understanding of translation**

Question 4. Given below is a single-stranded nucleotide sequence. Answer questions which refer to this sequence.

a. In the circle at the bottom of this sequence, place a 5' or 3', whichever corresponds.

b. Is the above structure an RNA or DNA? State which_____.

c. Assume that a complementary strand is produced in which all the innermost phosphates of the adenine triphosphonucleotide precursors are labeled with ^{32}P. What bases would be labeled if the complementary strand is completely degraded with spleen diesterase (cleaves between the phosphate and the 5'carbon)?_____.

d, What bases would be labeled if the complementary strand was completely degraded with snake venom diesterase (cleaves between the phosphate and the 3' carbon)?_____.

Concepts:

DNA and RNA structure

complementarity

5', 3' orientations

degradation products

Answer 4. (a) In this type of drawing, the various carbons of the sugar are readily apparent. It is the orientation and carbon numbering on the sugar which determines the 5'-3' orientation of the molecule. The bottom of the polymer has the 2' and 3' carbons projecting, while the top has the 5' carbon projecting. Therefore the bottom, near the circle is the 3' end of the molecule.

(b) Notice that there is no vertical line protruding from the 2' carbon position in the drawing and that uracil (U) is present. The molecule must therefore be an RNA. **(c)** It is best to start this portion of the problem by roughly drawing the complementary strand, remembering that it will be antiparallel.

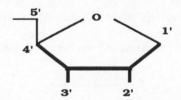

(c) If it is a DNA complement, it will have thymine in place of uracil. There is no indication as to the complementary strand being RNA or DNA. Since all ATP's (or dATP's) have at their innermost phosphate a ^{32}P, make certain that they are properly labeled as given below. Since spleen diesterase cleaves between the phosphate and the 5' carbon, the 5' neighbors (C, T or U) will be labeled with the ^{32}P.

(d) Since snake venom diesterase cleaves at the 3' position (between the phosphate and the 3' carbon), the originally labeled ATP (or dATP) will retain the label.

Common errors:

    5' to 3' orientations

    labeling of complementary strand

    understanding enzyme cleavages

Question 5. Assume that you were able to culture a strain of *E. coli* in medium containing either "normal" nitrogen or a heavy isotope of nitrogen (^{15}N). You grow the bacteria for a time in ^{15}N-containing medium which permits one complete replication of the bacterial chromosome. You extract the DNA, calling this extraction A. You continue to grow the bacterial culture in the ^{15}N DNA for a time which permits one more complete round of chromosome replication. You again extract the DNA, calling this extraction B. Assuming that non-labeled DNA has a density of 1.6 and that fully labeled DNA (that is with **all** the ^{14}N replaced with ^{15}N) has a density of 1.9, construct sedimentation profiles which reflect the expected densities of DNA from extractions A and B.

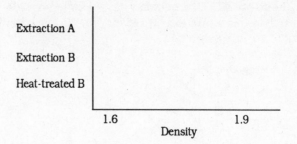

Knowing that heating DNA to 100° C. causes separation of complementary strands, use a broken line (- - -) to indicate the sedimentation profile of heat denaturation of extraction B DNA.

Concepts:

    semiconservative replication

    labeling

    centrifugation

    denaturation

Answer 5. Even though circular, in the context of this question, DNA from *E. coli* can be viewed in the following manner. Replication will occur semiconservatively and give the following sedimentation profile.

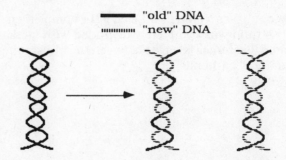

The heat treatment will cause the double-stranded structures to separate, giving the following strands and the profile as shown above.

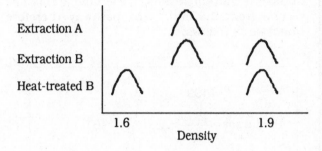

Extraction A

Extraction B

Heat-treated B

1.6   1.9

Density

Common errors:

   confusion with the labeling experiment

   difficulty in seeing sedimentation profiles

   application of semiconservative replication

Question 6. Assume that the following sequence of amino acids occurs in a protein starting from the "N" terminus (with asp) of a large polypeptide chain:

   asp-glu-ile-leu-ser-thr-met-arg-tyr-try-phe-gly

Assume that gene *X* is responsible for synthesis of this gene. Answer the questions below.

a.   Which amino acid(s) would you expect to change if gene *X* is altered by the mutagen, 2-amino purine, such that a transition mutation occurred which caused a change in the 9th base of the mRNA (counting from the 5' end of the coding region)?

b.   Which amino acid(s) would you expect to change if gene *X* is altered by the mutagen, acridine orange, such that a frameshift mutation occurred which caused an insertion of a base between bases 3 and 4 of the mRNA (counting from the 5' end of the coding region)?

c.   Which amino acid(s) would you expect to change if gene *X* is altered by the mutagen, nitrous acid, such that a mutation occurred which caused a change in the 11th base of the mRNA (counting from the 5' end of the coding region)?

Concepts:

   translation

   coding

   mutation

Answer 6. One of the most frequent difficulties students have with this problem is remembering that the code is triplet and three bases in the mRNA code for each amino acid in a protein. The 5' end of the mRNA corresponds with the N-terminus of the polypeptide chain.

(a) Transition mutations will cause amino acid substitutions. Counting over by "threes" from the N-terminus, the amino acid *ile* should be altered.

(b) Inserting a base between positions 3 and 4 will change the second amino acid; however, recall that acridine orange is a frameshift mutagen and insertion of a base will alter the reading frames for all "downstream" amino acids. Therefore all amino acids in positions two (glu) through twelve will be influenced (excluding degeneracy).

(c) Nitrous acid causes base substitutions, therefore a mutation in the 11th base would influence the amino acid *leu.*

Common errors:

  counting amino acids as bases

  **not understanding what base changes
  do to amino acid sequences**

Question 7. Below is a set of experimental results relating the growth (+) of *Neurospora* on several media. Based on the information provided, present the biochemical pathway and the locations of the metabolic blocks.

Strain   **Medium**

	MM	MM+A	MM+B
t409	-	+	+
t410	+	+	+
r3	-	-	+

Concepts:

  pathway analysis

  Beadle and Tatum "set-up"

  biochemical (nutritional) phenotypes

Answer 7. Notice that there are two mutant strains (can not grow on minimal medium) and one wild type strain (t410). The best way to approach these types of problems, especially when the data are organized in the form given, is to realize that the substance (supplement) which "repairs," as indicated by a (+), a strain is after the metabolic block for that strain. In addition, and most importantly, the substance which "repairs" the highest number of strains either *is the end product* or is *closest to* the end product.

Looking at the table, notice that supplement *B* "repairs" both the mutant strains. Therefore it must be at the end of the pathway or at least after all the metabolic blocks (defined by each mutation). Supplement *A* "repairs" the next highest number of mutant strains (1) therefore it must be second from the end. The pathway therefore would be as follows:

$$\text{Precursor}\text{----}\overset{\text{t409}}{\diagdown}\text{---->A}\text{----}\overset{\text{r3}}{\diagdown}\text{---->B}$$

To determine the locations at which the strains block, through mutation, the pathway, apply a similar logic. A strain that is "repaired" by all the supplements must be early in the pathway. A strain which is "repaired" by only one supplement must be late in the pathway. A supplement which does not "repair" a strain is before that strain's metabolic block.

Common errors:

  inability to construct a pathway

  **failure to see how additives "rapair"
  mutant phenotypes**

  **difficulty in assigning metabolic
  blocks in pathways**

Question 8. Prokaryotes and eukaryotes have evolved different mechanisms for obtaining more than one kind of protein from a single transcription unit (a transcription unit simply being a stretch of DNA that is transcribed into a single primary RNA transcript). Describe two different mechanisms.

Concepts:

overlapping genes

differential hnRNA splicing

Answer 8. In prokaryotes, overlapping genes present a mechanism for providing two and sometimes more protein products from a single stretch of DNA. In eukaryotes, different sets of introns may be removed, thus providing for a variety of protein products from a single section of DNA. This process is often called *differential splicing.*

Common errors:

Students often have difficulty in orienting *specific information* they have learned to a general question. If asked about overlapping genes, or differential hnRNA splicing, they would be able to develop an answer. Students sometimes confuse overlapping genes with the non-overlapping code.

Question 9. Drawn below is a hypothetical protein which contains areas where various types of bonds might be expected to occur. For each area a circle is drawn and in that circle, is placed a number. In the corresponding spaces below, state which bond type (or interaction) is most likely illustrated **and** state how that particular type of bond (or interaction) is formed. *You may use a given bond type only once.*

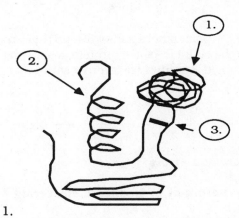

1.
2.
3.
4. What type of amino acids tend to be located on the outside (water side) of the molecule?

Concepts:

importance of primary structure

varieties of bonds

"higher level" folding

structure/function relationships

Answer 9.

1. Hydrophobic cluster formed by interaction of hydrophobic amino acids.

2. α helix formed from hydrogen bonds between components of the peptide linkage.

3. Covalent, disulfide bonds formed between cysteine residues

4. The polar amino acids will tend to orient to the outside of the protein where the charged R groups will interact with water.

Common errors:

    nature of hydrophobic clustering

    understanding of α and β structures

    **polar side chains and hydrophilic
        interactions**

Question 10. Drawn below is a diagram (not to scale) of DNA in the process of replication. Numbered arrows point to specific structures which you are to identify in the corresponding spaces below:

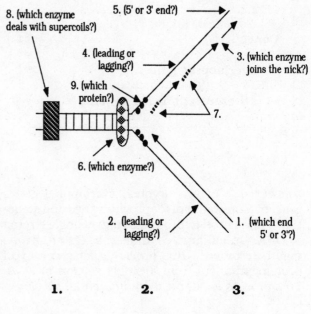

8. (which enzyme deals with supercoils?)

5. (5' or 3' end?)

4. (leading or lagging?)

3. (which enzyme joins the nick?)

9. (which protein?)

7.

6. (which enzyme?)

2. (leading or lagging?)

1. (which end 5' or 3'?)

1.   2.   3.

4.   5.   6.

7.   8.   9.

Concepts:

    overall DNA replication

    5', 3' polarity restrictions

    enzymology

    priming

Answer 10.

1. This must be the *5' end* of the polymer because all synthesis of polymers is 5' to 3' and the head of the arrow is at the other end of the polymer.

2. The *leading strand* is that strand which is synthesized continuously.

3. A *DNA ligase* will join the nicks.

4. The *lagging strand* is the discontinuous strand.

5. The free end that is complementary to the 5' end at arrow #1 must be the 3'end. That being so, the complement to that 3' end would be 5'. Therefore the *5' end* is at arrow #5.

6. A *helicase* is involved in unwinding the DNA helix.

7. An *RNA primer* is synthesized to initiate DNA synthesis.

8. *DNA gyrase* functions to remove supercoils generated by unwinding the DNA helix.

9. *Single-stranded binding proteins* stabilize the template which is to be replicated.

Common errors:

   determination of 5', 3' polarity

   naming of enzymes involved

15

Genetics of Bacteria and Viruses

Vocabulary: Organization and Listing of Terms

Structures and Substances

Salicin

Petri dish

    minimal medium

    liquid culture

    lawn of bacteria

Donor strain

    F sex pilus

    fertility factor, F factor

    tra

    Hfr

        circular chromosome

        operon

    F ', merozygotes

        partial diploid

rec genes, *A, B, C, D*

    D loop

Plasmid

    F factors

    R plasmids

        resistance transfer factor (RTF)

        r-determinants

        antibiotic resistance

    Col plasmids

        ColE1

            colicins

            colicinogenic

Insertion sequences (IS)

    perfect inverted repeat

Transposons (Tn)

    transposase enzyme

    bacteriophage *mu*

Heteroduplex

5-Hydroxymethylcytosine

Protein capsid

Lysozyme

Episome

Prophage P22

Concatemer

cos sites

   cosmids

RNA replicase

Reverse transcriptase

T lymphocyte

Pneumocystitis carini

Kaposi's sarcoma

Temperature sensitive conditional lethals

Processes/Methods

Spontaneous vs. Adaptive mutations (F15.1)

   lac⁻, lac⁺, val^R

   fluctuation test

Pure cultures

   prototroph

   auxotroph

   lag phase

   log phase

   stationary phase

   serial dilution

   sensitive

   resistant

   replica plating

Phenotypic and segregation lag

Bacterial recombination

   conjugation

      F⁺, F⁻

      physical contact

      unidirectional

      donor, "male"

      recipient, "female"

      fertility factor, F factor

      high frequency recombination, Hfr

         oriented transfer

         interrupted mating technique

         circular map

F' state

   merozygotes

   transformation

      entry

      recombination

      competence

   cotransformation

      linkage

   transduction

phage life cycle

latent (eclipse) period

    immediate early genes

    delayed early genes

    late genes

assembly

plaque (plaque assay)

    lysis

    virulent

lysogeny

    symbiotic relationship

    prophage

    temperate phage

    lysogenic bacterium

    vegetative

    U-tube experiment

        filterable agent (FA)

        prophage P22

    generalized transduction (F15.2)

        abortive transduction

        complete transduction

mapping

    specialized transduction

        prophage λ

            gal, bio

    cotransduced

Mutations (viral)

    rapid lysis, host range

    mixed infection experiments

    negative interference

    intergenic exchanges

    intragenic exchanges

Packaging

    concatemer

        headful mechanism

        circularly permuted

        terminally redundant

        phage λ

            cos sites

Viral reproduction

    bacteriophage ϕX174

        (+) strand

        (-) strand

        replicative form (RF)

        rolling circle mode

    bacteriophage QB

    poliovirus

    retrovirus

        reverse transcriptase

        RNA-directed DNA polymerase

oncogenic

Rous sarcoma virus (RSV)

provirus

transformed

human immunodeficiency virus (HIV)

    T lymphocyte

Recombinant DNA research

Transposition

  illegitimate recombination

    conservative

    replicative

Concepts

Spontaneous mutation

Directed mutation

  Adaptive mutations (F15.1)

    lac ⁻ —> *lac*⁺

    growth on salicin

    E. coli/T1 bacteriophage

Bacterial recombination

  relationship to *rec* genes

Complementation analysis

  complementation group

F15.1. Illustration of the expected fluctuation in colony number resulting from random mutations (spontaneous). Had the mutations "adapted" from some stimulus, if uniformly applied, the colony number would be more uniform from plate to plate.

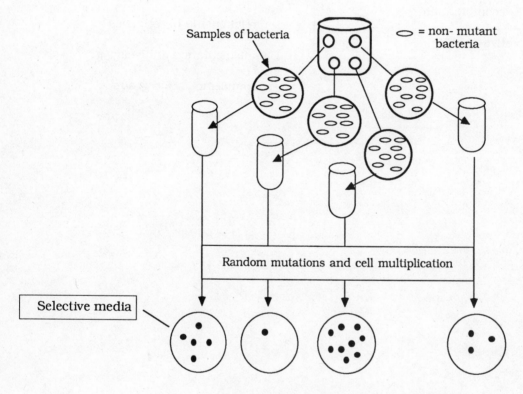

Fluctuation in number of mutant colonies

F15.2. Two forms of *generalized transduction* are depicted below. In *abortive transduction* the entering DNA does not integrate into the bacterial chromosome. In *complete transduction*, the entering DNA integrates into the bacterial chromosome.

Generalized Transduction

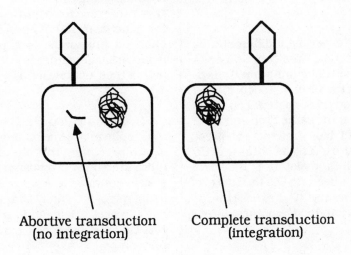

Abortive transduction
(no integration)

Complete transduction
(integration)

Solutions to Problems and Discussion Questions

1. The two major hypotheses concerning the origin of mutations in bacteria are *directed mutation* and *spontaneous mutation*. Directed mutation refers to genetic variations which are in response to an environmental stimulus. An example is the frequency of *lac⁻* to *lac⁺* changes in response to lactose in the medium. *Spontaneous mutations* are those which occur at random. Some spontaneous mutations will, by chance, be adaptive perhaps hundreds of generations later, others will have a negative impact on the organisms. It appears as if in some cases, for some genes, metabolic/genetic circumstances have evolved to provide for directed mutation.

2. The Luria-Delbruck experiment differentiates between "adaptive mutation" and "spontaneous mutation" on the premise that *if* mutational adaptation to the presence of the T1 phage occurs, then upon exposure to T1 a *group* of bacterial cultures will acquire mutations at the same time and have a similar frequency of T1-resistant bacteria present. However, if exposure to T1 has nothing to do with the acquisition of T1 resistance, then the frequency of T1 resistant bacteria will fluctuate in a *group* of cultures. Data (K/C-Tab.15.1) supported the *spontaneous mutation* hypothesis. See F15.1 in this book.

3. In replica plating a velvet pad is used to transfer bacteria, efficiently and free of media, from one Petri plate to another (K/C-Fig. 15.3). It is possible therefore to transfer a *pattern* of colonies from one type of *medium* to another. One can easily test for a variety of nutritional requirements (or antibiotic sensitivity or resistance) by transferring bacteria to a series of different Petri dishes containing various media.

4. Careful examination of K/C-Fig.15.4 shows that the *distribution* or pattern of T1 resistant bacteria was present *at the time*, and clearly before exposure to T1 phage, the velvet pad was inoculated with bacteria. Had T1 resistance been conferred at the time of T1 exposure, then the *patterns* of the colonies on the three plates in the figure would have varied (fluctuated). Using replica plating the question of *adaptive* vs. *spontaneous mutation* was easily answered.

5. *Phenotypic lag* relates to the time, in generations, that it takes to "dilute" the gene products from earlier generations. Proteins, even RNAs take some time to be diluted from a parental cell to the genetically altered progeny, therefore before a changed phenotype is expressed, the products from earlier generations must be diluted to a point where their expression is negligible.

Segregation lag results when the chromosomal replication time is shorter than the cell division time. Under this circumstance, there can be more than one chromosome in a bacterial cell. If a cell contains chromosomes of two different genotypes, it is likely that only one genotype will be expressed. Segregation of the chromosome will eventually yield a cell expressing the alternate phenotype.

Because mutations can occur and not be phenotypically expressed because of phenotypic or segregational lag, mutation rates may be *underestimated*.

6. Three modes of recombination in bacteria are *conjugation, transformation,* and *transduction.* Conjugation is dependent on the F factor which, by a variety of mechanisms, can direct genetic exchange between two bacterial cells. Transformation is the uptake of exogenous DNA by cells. Transduction is the exchange of genetic material using a bacteriophage.

7. (a) The requirement for physical contact between bacterial cells during conjugation was established by placing a filter in a U-tube such that the medium can be exchanged but the bacteria can not come in contact. Under this condition, conjugation does not occur.

(b) By treating cells with streptomycin, an antibiotic, it was shown that recombination would not occur if one of the two bacterial strains was inactivated. However, if the other was similarly treated, recombination would occur. Thus, directionality was suggested, with one strain being a donor strain and the other being the recipient.

(c) An F⁺ bacterium contains a circular, double-stranded, structurally independent, DNA molecule which can direct recombination.

8. (a) In an F⁺ X F⁻ cross, the transfer of the F factor produces a recipient bacterium which is F⁺. Any gene may be transferred, and the frequency of transfer is relatively low. Crosses which are Hfr X F⁻ produce recombinants at a higher frequency than the F⁺ X F⁻ cross. The transfer is oriented (non-random) and the recipient cell remains F⁻.

(b) Bacteria which are F⁺ possess the F factor, while those that are F⁻ lack the F factor. In Hfr cells the F factor is integrated into the bacterial chromosome and in F' bacteria, the F factor is free of the bacterial chromosome yet possesses a piece of the bacterial chromosome.

9. Mapping the chromosome in an Hfr X F⁻ cross takes advantage of the oriented transfer of the bacterial chromosome through the conjugation tube. For each F type, the point of insertion and the direction of transfer are fixed, therefore breaking the conjugation tube at different times produces partial diploids with corresponding portions of the donor chromosome being transferred. The length of the chromosome being transferred is contingent on the duration of conjugation, thus mapping of genes is based on time.

10. One can approach this problem by lining -up the data from the various crosses in the following order:

Hfr Strain	Order
1	T CHRO >>
2	HROM B >>
3	<< CHROM
4	M BAKT>>
5	<< BAKTC

Notice that all of the genes can be linked together to give a consistent map and that the ends overlap, indicating that the map is circular. The order is reversed in two of the crosses indicating the orientation of transfer is reversed.

11. In an Hfr X F⁻ cross, the F factor is directing the transfer of the donor chromosome. It takes approximately 90 minutes to transfer the entire chromosome. Because the F factor is the last element to be transferred and the conjugation tube is fragile, the likelihood for complete transfer is low.

12. As shown in K/C-Fig. 15.14, the F⁺ element can enter the host bacterial chromosome and upon returning to its independent state, it may pick up a piece of a bacterial chromosome. When combined with a bacterium with a complete chromosome, a partial diploid, or merozygote, is formed.

13. Transformation requires *competence* on the part of the recipient bacterium, meaning that only under certain conditions are bacterial cells capable of being transformed. Transforming DNA must be *double-stranded* to begin with yet is converted to a single-stranded structure upon insertion into the host cell. The most efficient length of the transforming DNA is about 1/200 of the size of the host chromosome. Transformation is an energy-requiring process and the number of sites on the bacterial cell surface is limited.

14. In the first data set, the transformation of each locus, a^+ or b^+, occurs at a frequency of .031 and .012 respectively. To determine if there is linkage one would determine whether the frequency of double transformants a^+b^+ is greater than that expected by a multiplication of the two independent events. Multiplying .031 X .012 gives .00037 or approximately 0.04%. From this information, one would consider no linkage between these two loci. Notice that this frequency is approximately the same as the frequency in the second experiment, where the loci are transformed independently.

15.

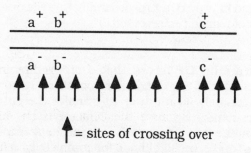

= sites of crossing over

Notice that the incorporation of loci a^+ and b^+ occurs much more frequently than the incorporation of b^+ and c^+ together (210 to 1) and the incorporation of all three genes $a^+b^+c^+$ occurs relatively infrequently. If a and b loci are close together and both are far from locus c then fewer crossovers would be required incorporate the two linked loci compared to all three loci. If all three loci were close together then the frequency of incorporation of all three would be similar to the frequency of incorporation of any two contiguous loci which is not the case.

16. In their experiment a filter was placed between the two auxotrophic strains which would not allow contact. F-mediated conjugation requires contact and without that contact, such conjugation can not occur. The treatment with DNase showed that the filterable agent was not naked DNA.

17. A *plaque* results when bacteria in a "lawn" are infected by a phage and the progeny of the phage destroy (lyse) the bacteria. A somewhat clear region is produced which is called a plaque.

Lysogeny is a complex process whereby certain temperate phage can enter a bacterial cell and instead of following a lytic developmental path, integrate their DNA into the bacterial chromosome. In doing so, the bacterial cell becomes lysogenic. The latent, integrated phage chromosome is called a *prophage*.

18. In *generalized transduction* virtually any genetic element from a host strain may be included in the phage coat and thereby be transduced. In *specialized transduction* only those genetic elements of the host which are closely linked to the insertion point of the phage can be transduced. Specialized transduction involves the process of lysogeny.

Because only certain genetic elements are involved in specialized transduction, it is not useful in determining linkage relationships. Cotransduction of genes in generalized transduction allows linkage relationships to be determined.

19. The first problem to be solved is the gene order. Clearly, the parental types are

$$a^+b^+c^+ \text{ and } a^-b^-c^-$$

because they are the most frequent. The double crossover types are the least frequent,

$$a^-b^-c^+ \text{ and } a^+b^+c^-.$$

Because it is the gene in the middle which switches places when one compares the parental and double crossover classes, the *c* gene must be in the middle. The map distances are as follows:

$$a \text{ to } c = (740 + 670 + 90 + 110)/10,000$$

$$= 16.1 \text{ map units}$$

$$c \text{ to } b = (160 + 140 + 90 + 110)/10,000$$

$$= 5 \text{ map units}$$

To determine the type of interference, first determine the *expected* frequency of double crossovers ($0.161 \times .05 = .000805$), which when multiplied by 10,000 gives approximately 80. The *observed* number of double crossovers is $90 + 110$ or 200. Since many more double crossovers are observed than expected, negative interference is occurring.

20. Viral recombination occurs when there is a sufficiently high number of infecting viruses so that there is a high likelihood that more than one type of phage will infect a given bacterium. Under this condition phage chromosomes can recombine by crossing over as indicated in K/C-Fig.15.28.

21. As shown in K/C-Fig.15.28, crossing over produces multiple genomes, called concatemers, in single DNA strands. During maturation of the phage, the concatemers are enzymatically cleaved into pieces which are large enough to fill the head capsid (headful packaging).

Because the pieces are slightly larger than a single genome, each DNA strand is terminally redundant. In addition, because the cleavage sites generate terminally redundant ends, the precise cut sites of the endonuclease occurs at different points along the concatemer thus generating genomes which are circularly permuted.

22. The approach for determining the complementation groupings and the results of the missing data is to recall that if a "+" is registered, different complementation groups (genes) exist. If a "-" results, then the two mutations are in the same complementation group. For Experiment #1, *d* and *f* are in the same complementation group (gene) while *e* is in a different gene. Therefore

$$e \times f = +.$$

For Experiment #2, all three mutations are in the same gene, hence

$$h \times i = -.$$

In Experiment #3 we learn that *d* and *g* are in different genes, while *e* and *i* are in the same gene. Putting all of the data together, we see that there are two genes: *d* and *f* are in the same gene while *e*, *g*, *h*, and *i* are in a different gene.

23. (a) Culture #1 represents a cell concentration of 240 cells/ml X 10^9 (recall that only 0.1ml of the various dilutions was plated) which had been irradiated. Because leucine is added to the minimal medium, both *leu⁺* and *leu⁻* cells will grow.

Culture #2 represents a cell concentration of 120 cells/ml X 10^2 which had been irradiated. Because leucine is not added to the minimal medium, only *leu⁺* cells will grow.

Culture #3 represents a cell concentration of 120 cells/ml X 10^9 which had not been irradiated. Because leucine is added to the minimal medium, both *leu⁺* and *leu⁻* cells will grow.

Culture #4 represents a cell concentration of 30 cells/ml X 10^1 which had not been irradiated. Because leucine is not added to the minimal medium, only *leu⁺* cells will grow. One would expect the values to be similar in cultures #1 and #3 because, with leucine added to the medium, one can not differentiate between *leu⁺* and *leu⁻* cells.

However, it is likely that new non-leucine related nutritional mutations might be induced by the irradiation.   If anything therefore, we might expect to see fewer colonies from culture #1 when compared to #3.   The difference of **240 cells/ml** X 10^9 and **120 cells/ml** X 10^9 may be within the limits of experimental error.

(b) The general formula for determining mutation rate is to divide the number of mutant bacteria by the total number of bacteria. In this experiment the spontaneous mutation rate would be calculated as follows:

$$(30 \times 10^1)/(120 \times 10^9) = 0.25 \times 10^{-8}$$

The induced mutation rate would be calculated as follows:

$$(120 \times 10^2)/(240 \times 10^9) = 0.5 \times 10^{-7}$$

24. Insertion sequences are relatively short sections of DNA which contain genetic information related to the insertion process. Transposons on the other hand are larger sections of DNA which contain insertion sequences which are covalently linked to other genes, notably those involved in antibiotic resistance.

25. Starting with a single bacteriophage, one lytic cycle produces 200 progeny phage, three more lytic cycles would produce $(200)^4$ or 1,600,000,000 phage.

16

Recombinant DNA

Vocabulary: Organization and Listing of Terms

Structures and Substances

Restriction endonucleases

    Eco R1

Vector

    cloning vehicle

        plasmids

        bacteriophage

        cosmids

Cloned DNA fragments

Modification enzymes

Poly dA

Poly dT

Reverse transcriptase

Ribonuclease H

DNA polymerase I

S_1 nuclease

Nitrocellulose

Ethidium bromide

X-gal

Open reading frame

Opines

Callus

Amnionic fluid

Allele specific oligonucleotide

Processes/Methods

Recombinant DNA technology

    gene splicing

    genetic engineering

        restriction endonucleases

            type I

            type II

                palindromic sequences

sticky ends

modification enzymes

terminal deoxynucleotidyl
transferase

poly dA

poly dT

vector

plasmids

bacteriophage

transfection

M13

DNA sequencing

cosmids

cos sequences (lambda)

shuttle vector

cloned DNA fragments

autonomous replication

amplification

antibiotic resistance

ampicillin resistance

tetracycline resistance

hosts

E. coli K12

transformation

yeast

2 micron plasmid

integrative plasmid

integrative transformation

yeast artificial chromosome

plants

Agrobacterium tumifaciens

tumor-inducing plasmid

T-DNA

callus

transgenic

mammals

calcium phosphate

endocytosis

microinjection

electroporation

encapsulation/fusion

vectors

bovine papilloma virus

retroviruses

transgenic

library construction

genomic libraries

$N = \ln(1\text{-}P)/\ln(1\text{-}f)$

chromosome-specific libraries

    subgenomic fraction

    flow cytometry

    pulse field gel electropohoresis

cDNA libraries

    reverse transcriptase

    ribonuclease H

    DNA polymerase I

    S_1 nuclease

selection of recombinant clones

    probes

    reverse translation

colony and plaque hybridization

mixed ligation products

    X-gal

        blue, clear plaques

    nitrocellulose

PCR analysis

    denaturation

    annealing of primers

    extension of primers

restriction mapping

    restriction digestion

    gel electrophoresis

ultraviolet illunimation

nucleic acid blotting

    Southern blot

        autoradiography

    Northern blot

    densitometer

    Western blot

DNA sequencing

    Maxim and Gilbert

    Sanger

applications

    gene mapping

        RFLPs

        multigenerational families

    disease diagnosis

        amniocentesis

        chorionic villus sampling

    commercial applications

Concepts

Restriction mapping

Gene mapping

Solutions to Problems and Discussion Questions

1. Recombinant DNA technology, also called genetic engineering or gene splicing, involves the creation of associations of DNA that are not typically found in nature. Particular enzymes, called *restriction endonucleases*, cut DNA at specific sites and often yield "sticky" ends for additional interaction with DNA molecules cut with the same class of enzyme.

Isolated from bacteria, restriction enzymes fall into two classes, type I and type II. *Terminal transferase* is an enzyme which extends single-stranded ends with the addition of "tails" which provide "sticky" ends. A *vector* may be a plasmid, bacteriophage, or cosmid which receives, through ligation, a piece, or pieces, of foreign DNA. The recombinant vector can transform (or transfect) a host cell (bacterium, yeast cell, etc.) and be amplified in number. *Calcium chloride* is often used to increase the permeability of host cells to transformation.

2. *Reverse transcriptase* is often used to promote the formation of cDNA (complementary DNA) from a mRNA molecule. Eukaryotic mRNAs typically have a 3' polyA tail as indicated in the diagram below. The poly dT segment provides a double-stranded section which serves to prime the production of the complementary strand.

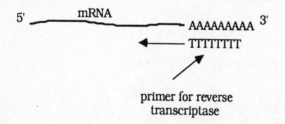

3. **(a)** Because the *Drosophila* DNA has been cloned into the *Eco R1* site in the ampicillin resistance gene of the plasmid, the gene will be mutated and any bacterium with the plasmid will be ampicillin sensitive. The tetracycline resistance gene remains active however. Bacteria which have been transformed with the recombinant plasmid will be resistant to tetracycline and therefore tetracycline should be added to the medium.

(b) Colonies which grow on a tetracycline medium should be tested for growth on an ampicillin medium either by replica plating or some similar controlled transfer method. Those bacteria which do not grow on the ampicillin medium probably contain the *Drosophila* DNA insert.

(c) Resistance to both antibiotics by a transformed bacterium could be explained in several ways. First, if cleavage with the *Eco R1* was incomplete, then no change in biological properties of the uncut plasmids would be expected. Also, it is possible that the cut ends of the plasmid were ligated together in the original form with no insert.

4. Apply the formula:

$$N = \ln(1-P)/\ln(1-f)$$

$$= \ln(1-0.99)/\ln(1-[5 \times 10^3/1.5 \times 10^8])$$

$$= \ln(.01)/\ln(.9999667)$$

$$= -4.605/-0.0000333$$

$$= 1.38 \times 10^5$$

5. When the *insulin* gene is "manufactured" it is made as a complementary copy of the mRNA, which is void of introns. Therefore the insulin mRNA which is made from the cDNA does not have introns.

6. The question of protein/DNA recognition and interaction is a difficult one to answer. Much research has been done to attempt to understand the nature of the specificity of such interactions. In general it is believed that the protein interacts with the major groove of the DNA helix. This information comes from the structure of the few proteins which have been sufficiently well studied to suggest that the DNA major groove and "fingers" or extensions of the protein form the basis of interaction.

7. Given that there is only one site for the action of *Hind*III, then the following will occur. Cuts will be made such that a four base single-stranded set of sticky ends will be produced. For the antibiotic resistance to be present, the ligation will reform the plasmid into its original form. However, two of the plasmids can join to form a dimer as indicated in the diagram below.

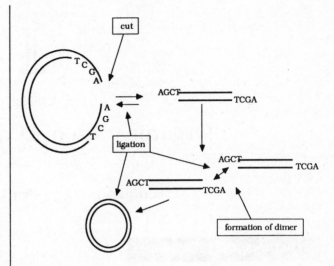

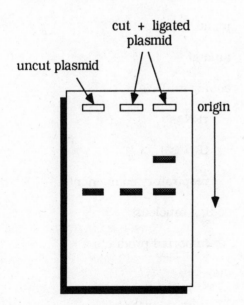

8. The regulated output of a gene, in balance with other gene products in a cell, is necessary for normal physiological function. Regulation is achieved at several levels from cell-to-cell interactions to position effects caused by neighboring genes. The overall positional, temporal, and molecular environment of a gene, about which there is often little information, will ultimately determine the effectiveness, in a therapeutic sense, of any introduced gene. Before gene therapy is to be dependable such variables will need to be understood. Even though a great deal is known about the expression of hemoglobin genes, the information presented in the question indicates tissue-specific influences which cause the imbalance leading to thalassemia.

9. All other factors being equal (appropriate cloning sites and selectable markers), it is important to consider the size of the foreign DNA which can cloned into the vector. Generally for large genomes, it is best to use a vector which will accept relatively large fragments.

10. Even though there is great economic potential in achieving sophisticated genetic engineering in plants, there are relatively few cloning vectors available for economically important crops.

17

Organization of DNA in Chromosomes

Vocabulary: Organization and Listing of Terms

Structures and Substances

Viral chromosomes

  DNA, RNA

    double-stranded

    single-stranded

    often circular

  capsid

  φX174

  polyoma

  lambda (λ)

    polynucleotide ligase

  T-even bacteriophages

  tobacco mosaic virus (TMV)

Bacterial chromosomes

  DNA

  double-stranded

  nucleoid

E. coli

  circular

  DNA-binding proteins

    HU, H

Mitochondrial DNA (mtDNA)

  plant

  animal

  coding (mtDNA)

    rRNAs

    tRNAs

    respiratory components

  coding (nuclear)

    imported products

  diversity

    human mtDNA

    yeast mtDNA

      introns

Chloroplast DNA (cpDNA)

    circular

    double-stranded

    different than nuclear DNA

    coding (cpDNA)

        rRNAs

        tRNAs

        ribulose-1-5-biphosphate carboxylase

Eukaryotic chromosomes

    chromatin

    mitotic chromosomes

        condensed chromatin

            folded fiber

        sister chromatids

        karyotype

            Denver classification system

            chromosome bands

            chromosome mutations
                (aberrations)

        folded fiber

            nucleosome and solenoid

        synaptonemal complex

            central element

            lateral elements

            c(3)G mutation

Polytene chromosomes

    dipteran larval cells

    parallel register

    bands, interbands, chromomeres

    puffs

Lampbrush chromosomes

    linear axis

    lateral loops

        deoxyribonucleoprotein (DNP)

        one DNA helix

    central axis

        two DNA helices

Chromatin

    nucleoprotein

    histones

        amino acid composition

    nonhistones

    micrococcal nuclease

        ν-bodies, nucleosomes

            tetramers

            nucleosome core particle

        linker DNA

            histone H1

platysome

packing ratio

solenoid

Heterochromatin

unineme

telomeres

single-stranded nucleotide gaps

tandem repeats

telomere-associated sequences

telomerase

centromeres

spindle fibers

CENs

Regions I, II, III

kinetechore

chromocenter

constitutive

facultative

Barr body

mealy bug chromosomes

position effect

satellite DNA

repetitive DNA

tandem repeats

Processes/Methods

Maternal mode of inheritance

Semiconservative replication

Importing of nuclear-coded gene products

Exchange of nuclear/mtDNA

Folded-fiber (eukaryotic chromosome)

coiling-twisting-condensing

Complementation groups

Heterochromatin

few genes

late replicating

position effect

telomere replication

Molecular Analysis

Light microscopy

Electron microscopy

Autoradiography

Equilibrium density sedimentation

Chromosome banding

C-banding

Q-banding

G-banding

R-banding

Cytological mapping

    genetic (linkage) maps

    cytological maps

Scanning electron microscopy (SEM)

Transmission electron microscopy (TEM)

X-ray diffraction

Neutron scattering

Reassociation kinetics

In situ hybridization

Concepts

    Evolution of cellular organelles

        endosymbiont theory

        molecular diversity

        genetic code

        sensitivity to antibiotics

    One band-one gene paradigm

        evidence for and against

    Unineme model of chromosomes

    Heterochromatin/euchromatin

*Solutions to Problems and
Discussion Questions*

1. General similarities and differences:

Viral	*Bacterial*
DNA or RNA	DNA
single-stranded	double-stranded
double-stranded	circular (*E. coli*)
linear	DNA-binding proteins
ring shaped (circular)	genome size >100μm
naked nucleic acid	
genome size < 100μm	

2. General similarities and differences:

mtDNA	*cpDNA*
circular	circular
double-stranded	double-stranded
semiconservative repl.	semiconservative repl.
animal (16 to 18 kb)	>100 kb
plant (>100 kb)	genes(rRNAs, tRNAs, etc.)
genes (rRNAs, tRNAs, etc.)	
diverse (introns in some)	
variations in genetic code	

3. While greater DNA content per cell is associated with eukaryotes, one can not universally equate genomic size with an increase in organismic complexity. There are numerous examples where DNA content per cell varies considerably among closely related species. Because of the diverse cell types of multicellular eukaryotes, a variety of gene products is required, which may be related to the increase in DNA content per cell. In addition, the advantage of diploidy automatically increases DNA content per cell.

However, seeing the question in another way, it is likely that a much higher *percentage* of the genome of a prokaryote is actually involved in phenotype production that in a eukaryote.

Eukaryotes have evolved the capacity to obtain and maintain what appears to be large amounts of "extra" perhaps "junk" DNA. This concept will be examined in subsequent chapters. Prokaryotes on the other hand, with their relatively short life cycle, are extremely efficient in their accumulation and use of their genome.

Given the larger amount of DNA per cell and the requirement that the DNA be partitioned in an orderly fashion to daughter cells during cell division, certain mechanisms and structures (mitosis, nucleosomes, centromeres, etc.) have evolved for *packaging* the DNA. In addition, the genome is divided into separate entities (chromosomes) to perhaps facilitate the partitioning process in mitosis and meiosis.

4. Polytene chromosomes are typical of some cells of Dipteran larvae. They arise from a series of DNA replications without chromosome separation. Strands of DNA are in "parallel register." When viewed under the light microscope, polytene chromosomes appear *banded* as in K/C-Fig. 17.9. The dark regions are called *bands* (also called *chromomeres*) and the lighter regions between the bands are called *interbands*. Local unwinding of a section of a polytene chromosome in which RNA is being synthesized, is called a *puff*. Until fairly recently, one band was considered to be one gene.

5. Digestion of chromatin with endonucleases, such as micrococcal nuclease, gives DNA fragments of approximately 200 base pairs or multiples of such. X-ray diffraction data indicated a regular spacing of DNA in chromatin. Regularly spaced bead-like structures (nucleosomes) were identified by electron microscopy. Nucleosomes are octomeric structures of two molecules of each histone (H2A, H2B, H3, and H4) except H1. Between the nucleosomes and complexed with linker DNA is histone H1. A 146-base pair sequence of DNA wraps around the nucleosome and as chromosome condensation occurs a 300-A° fiber is formed. It appears to be composed of 5 or 6 nucleosomes coiled together. Such a structure is called a solenoid.

6. The *synaptonemal complex*, being a triplet of parallel strands (central element and lateral elements) probably provides stability to the synapsed chromosomes so that crossing over may occur. While not being required for pairing of homologous chromosomes, the synaptonemal complex greatly influences the frequency of crossing over. In zipperlike fashion it brings homologues into intimate association.

7. The endosymbiont theory states that mitochondria and chloroplasts evolved from free-living bacteria-like structures that entered a symbiotic relationship with a host cell. In time, both the bacteria-like particle and the host cell became interdependent through the loss and sharing of products. In addition to the size (sedimentation properties) of ribosomes, the ribosomes of mitochondria and chloroplasts show antibiotic sensitivity which is similar to present-day bacteria, thus supporting the endosymbiont theory.

8. *Heterochromatin* is chromosomal material which stains deeply and remains condensed when other parts of chromosomes, euchromatin, are otherwise pale and decondensed. Heterochromatic regions replicate late in S phase and are relatively inactive in a genetic sense because there are few genes present or if they are present, they are repressed. Telomeres and the areas adjacent to centromeres are composed of heterochromatin.

9. *Constitutive heterochromatin* includes regions at the same sites on homologous chromosomes which are genetically "inert" such as telomers and centromeric regions. *Facultative* heterochromatin has the potential to become heterochromatic such as the X chromosome in female mammals (Barr body) and one haploid set of chromosomes in the mealy bug.

10. The banding patterns on the polytene chromosomes which were discovered in the 1930s provided "landmarks" which allowed geneticists to consistently orient themselves to *specific* regions of chromosomes.

Banding allowed the early description of chromosomal aberrations and eventually led to the development of cytological maps as well as the evolution of karyotypes of related species.

11. (a) Since there are 200 base pairs per nucleosome (as defined in this problem) and 10^9 base pairs, there would be 5×10^6 nucleosomes.

(b) Given that there are 5 nucleosomes per solenoid, there would be 1×10^6 solenoids.

(c) Since there are 5×10^6 nucleosomes and nine histones (including H1) per nucleosome, there must be $9(5 \times 10^6)$ histone molecules: 4.5×10^7.

(d) Since there are 10^9 base pairs present and each base pair is 3.4 A° the overall length of the DNA is 3.4×10^9 A°. Dividing this value by the packing ratio (50) gives 6.8×10^7.

12. The first part of this problem is to convert all of the given values to cubic A° remembering that 1 µm = 1000 A°. Using the formula πr^2 for the area of a circle and $4/3 \pi r^3$ for the volume of a sphere, the following calculations apply:

Volume of DNA: $3.14 \times 10A° \times 10A° \times (50 \times 10^4 A°) = 1.57 \times 10^8 A°^3$

Volume of capsid: $4/3 (3.14 \times 400A° \times 400A° \times 400A°) = 2.67 \times 10^8 A°^3$

Because the capsid head has a greater volume than the volume of DNA, the DNA will fit into the capsid.

18

Organization and Structure of Genes

Vocabulary: Organization and Listing of Terms

Structures and Substances

Repetitive sequences

   highly repetitive

      Alu family

      mobile genetic elements

   middle or moderately repetitive

Multigene families

   globins

      α family

      β family

   histones

   ribosomal RNA

      tandem repeats

      nontranscribed spacer regions

      nucleolar organizer regions (NORs)

Cistron

Pseudoallele

   subloci

Xanthine dehydrogenase

Heterogeneous RNA (hnRNA)

   introns (F18.1)

   exons (F18.1)

      protein domains

Promoter regions

      Goldberg-Hogness or TATA box

      Pribnow sequence

      CAAT box

      enhancers

Intergenic regions

LDL receptor protein

Processes/Methods

Fine structure genetic analysis

    allelism

        similar phenotype

        similar location

    complementation

        cis, trans

            rII in T4 phage

                E. coli B

                E. coli K12

                    A, B cistrons

            Drosophila

                lozenge (lz)

                    complex locus

                    pseudoalleles (F18.2)

                    intergenic recombination

                    intragenic recombination

                    subloci (F18.2)

                rosy (ry)

                    purine selection

                bithorax gene complex

    exon shuffling (F18.1)

        LDL receptor protein

            endocytosis

            epidermal growth factor (EGF)

Recombinant DNA technology

Nucleotide sequencing technologies

Deletion mapping (testing)

Concepts

Genetic analysis vs. biochemical analysis

    fine structure genetic analysis

        allelism testing

        complementation testing

            cis, trans

    cistron

    mutational hot spots

    pseudoallele

C value paradox

    introns

    promoter regions

        flanking regions

    intergenic regions

Exon shuffling and protein domains (F18.1)

Evolution of multigene families

    globins

    histones

    ribosomal RNAs

F18.1. Illustration of the relationship between exons and protein domains. *Exon shuffling* is illustrated where exons from gene *A* are found in genes *B* and *C* which each have only one exon (#2 and #3, respectively) in common with gene *A*.

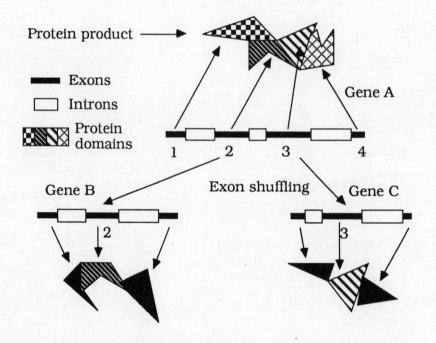

F18.2. Illustration of intergenic and intragenic recombination. *Pseudoalleles* occur when intragenic recombination separates subloci. *Subloci* are shown in the bottom portion where crossovers can not occur between mutation sites.

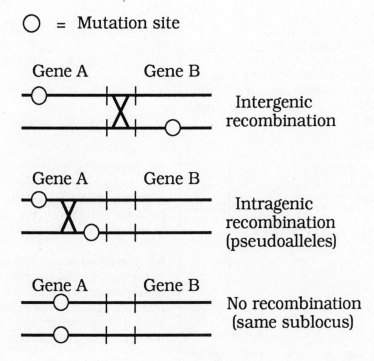

Solutions to Problems and Discussion Questions

1. The experiments are set up such that the *trans* configuration exists in the brief diploid phase. Thus if a "+" occurs, the two mutations *complement* and are in different genes (complementation groups). Merely make pairwise comparisons and place each combination with a "-" in the same group. Below is an analysis of the data as presented:

Three groups:

2. Mutations which are placed in the *cis* configuration will have a "+" phenotype while those which are in the same complementation group will show a "-" if in *trans*. Mutations which are in a different complementation group will show "+" if in *trans*. The data then would be as follows:

	Condition	
Mutants	*Cis*	*Trans*
1,2	+	-
1,3	+	+
2,3	+	+

3.

Group A: *e X f* = lysis because *d* and *e* are in different complementation groups and *d* and *f* are in the same cistron, therefore *e* and *f* must be in different cistrons.

Group B: *h X i* = no lysis because *g, h,* and *i* are in the same complementation group.

Group C: *k X l* = no lysis because mutations *jj* and *k* are in different cistrons and *j* and *l* are in different cistrons therefore genes *k* and *l* are in the same cistron.

4. Becaues there are only two complementation groups in the *rII* region one would have the following groupings:

    Group A: 1,4,5   *Group B:* 2,3

(a) Therefore the result of testing

    2 X 3 = no lysis;

    2 X 4 = lysis;

    3 X 4 = lysis.

(b) Because mutant 5 failed to complement with mutations in either cistron, it probably represents a major alteration in the gene such that both cistrons are altered. A deletion which overlaps both cistrons could cause such a major alteration.

(c) The recombination frequency is given by the following formula. Recall that only one of the two recombinant types are recovered in this type of experiment where the assay of growth on *E. coli* B is used. Remember to include the dilutin factor in the setting of the observed values.

General formula:

$$\frac{2(\text{number of recombinant types})}{\text{total number of progeny}} =$$

$$2(5 \times 10^1)/(2 \times 10^5) = 5 \times 10^{-4}$$

(d) Because mutant 6 complemented mutations 2 and 3, it is likely to be in the cistron with mutants 1,4, and 5. A lack of recombinants with mutant 4 indicates that mutant 6 is a deletion which overlaps mutation 4. Recombinants with 1 and 5 indicates that the deletion does not overlap these mutations.

5. A *pseudoallele* is a term used to describe an allele which by all standard criteria (complementation, mapping, phenotype) behaves as an allele to another gene but displays recombination with that gene. The *lozenge* gene contains pseudoalleles therefore it is viewed as a complex locus. See F18.2.

6. *Exon shuffling* is a term used to describe the likely phenomenon whereby the exons of genes, each coding for functional domains of proteins, can be "shuffled" or rearranged to facilitate the evolution of a new protein. Rather than each protein evolving "from scratch" it is suggested that they are built from previously evolved gene segments, exons. Several lines of evidence support the "exon shuffling" model:

(a) exon size is consistent with the size of functional domains of proteins,

(b) exon/intron boundaries often match up with functional domains of proteins,

(c) the major recombinational events occur in the introns, thus leaving the exons as potentially mobile units, and

(d) striking sequence homology exists between exons from genes which code for different proteins. See F18.1.

7. While the β-globin gene family is relatively large (60kb) sequence and restriction analyses show that it is composed of six genes, one is a pseudogene and therefore does not produce a product.

The five functional genes each contain two similarly-sized introns which when included with non-coding flanking regions (5' and 3'), and spacer DNA between genes, accounts for the 95% mentioned in the question.

8. The α and β globin families are well characterized and shown to exist at two locations in the human genome (chromosome #16 and #11, respectively). There are sequence homologies between the two and each shows significant regulation during development. The α gene family consists of about 30kb and contains 5 genes including two pseudogenes. Each gene contains two introns and the protein products of the functional genes are identical in length.

The β gene family contains six genes and occupies 60kb of DNA. A pseudogene is present. All five functional genes are 146 amino acids in length and each contains two similarly-sized introns.

The *histone* gene family or cluster contains five related, but non-identical genes which are separated by non-transcribed spacer sequences. Almost all histone genes lack introns as well as the 3'-flanking region which is involved in the addition of the poly A tail. The size of the histone gene complex is about 10kb which is smaller than the globin gene families.

9. The *rRNA* gene family is well characterized in a variety of organisms. It generally consists of tandem repeats of three molecules in the following order: 18S, 5.8S, and 28S rRNA. There are substantial regions between each gene which are transcribed but subsequently removed. Thus, the initial transcript is processed similar to the removal of introns. Between each three-gene unit in the cluster is a non-transcribed spacer DNA sequence.

The transcription unit varies in size from 7kb to 13kb in humans, the variation coming from the different sized spacer DNAs. The basic family unit, including the non-transcribed spacers contains about 43kb of tandemly repeated DNA.

In humans the rRNA gene clusters are located on the ends of chromosomes #13, #14, #15, #21, and #22 while in *Drosophila* they are located on the X and Y chromosomes. The chromosomal regions which include the DNA coding for rRNAs are called *nucleolus organizer regions* (NORs).

10. First a "C value" is the amount of DNA contained in a haploid genome. The *C value paradox* recognizes that with evolutionary divergence there has been a dramatic divergence in the amount of DNA among different taxa. It is likely that with the phylogenetic divergence there is not a commensurate requirement for *that many* different and additional genes. Indeed, some organisms which are quite closely related phylogenetically have vastly different DNA contents. So what is this "extra" DNA doing in various genomes?

Until recently, answers to this question were vague and speculative. However, with recent advances in molecular biology, especially cloning and sequencing of DNA, it has been determined than much of this "extra" DNA is found in heterochromatic regions, introns, regions flanking genes, and intergenic spacer regions. While the nature of large amounts of "non-coding" DNA has been described, the question that still remains is "why do different taxa retain such different amounts of this "extra" DNA?" The C value paradox is being answered at a "within taxon" level but not too well at the "among taxa" level.

19

Genetic Regulation in Bacteria and Viruses

Vocabulary: Organization and Listing of Terms

Structures and Substances

Lactose

    structural genes

        lac z

            β-galactosidase

        lac y

            β-galactoside permease

        lac a

            transacetylase

    polycistronic mRNA

    gratuitous inducers

        isopropylthiogalactoside (IPTG)

    constitutive mutants

        lac⁻

        lac O^c

    regulatory units

        repressor gene

            repressor molecule

            diffusible cellular

                product(F19.2)

        operator region

            no diffusible product

            adjacent control (F19.2)

    lac i^s

    lac i^a

  catabolite activating protein (CAP)

    glucose

    CAP binding site

    cyclic adenosine monophosphate (cAMP)

    adenyl cyclase

Arabinose

  ara B, A, D

  ara regulatory protein

Tryptophan

  tryptophan synthetase

  trp R, *trp R*⁺

  structural genes

    trp E, D, C, B, A

  trp P-trp O region

  leader sequence

    attenuator

Lambda DNA

  λ repressor protein

    cI repressor

      negative control element

  cro protein

  cII protein

  rec A protein

T7

  early genes

    viral-specific RNA polymerase

    protein kinase

Phage SPO1

  early genes

  middle genes

  late genes

Processes/Methods

Genetic regulation

  adaptive

    inducible

      lactose

    repressible

      tryptophan

        attenuation

          hairpin loop

          ribosome "stall"

  negative control

  positive control

    catabolite repression

    arabinose regulation

  constitutive

  allosteric

  λ lysogeny or lysis

    phage *lambda*

    SOS response

Phage transcription

   T7

Equilibrium dialysis

Ultracentrifugation

   glycerol gradient

Concepts

Genetic regulation

   efficiency

      activity of enzymes

      level of transcription

Positive (F19.1)

   catabolite repression

Negative control (F19.1)

   lactose

      induction

   tryptophan

      repression

Operon model

Lambda regulation

F19.1. Illustration of general processes of *negative* and *positive* control. In *negative* control, the regulatory protein inhibits transcription while under *positive* control, transcription is stimulated.

NEGATIVE CONTROL

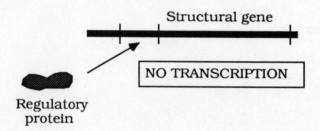

POSITIVE CONTROL

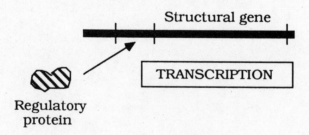

F19.2. Illustration of the nature of the product of the i gene. It can act "at a distance" because it is a protein which can diffuse through the cytoplasm and thus act in "trans" as well as in "cis." There is no protein product of the operator gene, therefore it can only act in "cis."

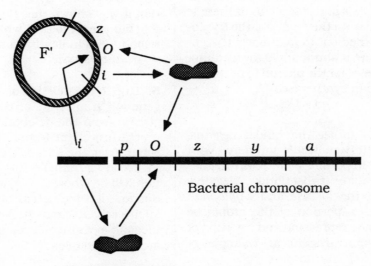

Bacterial chromosome

repressor protein

Solutions to Problems and Discussion Questions

1. The answer to this question is a key to enhancing a student's understanding of the Jacob-Monod model as related to lactose and tryptophan metabolism. The enzymes of the lactose operon are needed to break down and use lactose as an energy source. If lactose is the sole carbon source, the enzymes are synthesized *to use* that carbon source. With no lactose present, there is no "need" for the enzymes.

The tryptophan operon contains structural genes for the *synthesis* of tryptophan. If there is little or no tryptophan in the medium, the tryptophan operon is "turned on" to manufacture tryptophan. If tryptophan is abundant in the medium, then there is no "need" for the operon to be manufacturing "tryptophan synthetases."

2.   Refer to F19.1 to see that under *negative* control, the regulatory molecule interferes with transcription while in *positive* control, the regulatory molecule stimulates transcription. Negative control is seen in the *lactose* and *tryptophan* systems as well as a portion of the *arabinose* regulation. Catabolite repression and a portion of the *arabinose* regulatory systems are examples of positive control.

3. In an *inducible system*, the repressor which normally interacts with the operator to inhibit transcription, is inactivated by an *inducer*, thus permitting transcription. In a *repressible system*, a normally inactive repressor is *activated* by a *co-repressor*,   thus enabling it (the activated repressor) to bind to the operator to inhibit transcription. Because the interaction of the protein (repressor) has a negative influence on transcription, the systems described here are forms of *negative control* (see F19.1).

4. A single *E. coli* cell contains very few molecules of the *lac* repressor. However, the *lac* t^q mutation causes a 10X increase in repressor protein production, thus facilitating its isolation. With the use of dialysis against a radioactive gratuitous inducer (IPTG), Gilbert and Muller-Hill were able to identify the repressor protein in certain extracts of *lac* t^q cells. The material which bound the labeled IPTG was purified and shown to be heat labile and have other characteristics of protein. Extracts of *lac t* cells did not bind the labeled IPTG.

The IPTG-binding protein was labeled with sulfur-containing amino acids and mixed with DNA from λ phage which contained the *lac O⁺* section of DNA. By glycerol gradient centrifugation it was shown that the labeled repressor protein binds only to DNA which contained the *lac O⁺* region, thus indicating a *specific* binding to DNA.

5. Due to the deletion of a base early in the *lac z* gene will "frameshift" all the reading frames downstream from the deletion, it is likely that either premature chain termination of translation will occur (from the introduction of a nonsense triplet in a reading frame) or the normal chain termination will be ignored. Regardless, a mutant condition for all three structural genes will be likely. If such a cell is placed on a lactose medium, it will be incapable of growth because the needed enzymes are not available.

6. Refer to **K/C-Fig. 19.4** and to F19.2 to get a good understanding of the lactose system before starting.

$i^+ o^+ z^+$ = **Inducible** because a repressor protein can interact with the operator to turn off transcription.

$i^- o^+ z^+$ = **Constitutive** because the repressor gene is mutant, therefore no repressor protein is available.

$i^+ o^c z^+$ = **Constitutive** because even though a repressor protein is made, it can not bind with the mutant operator.

$i^-o^+z^+$ / F' i^+ = **Inducible** because even though there is one mutant repressor gene, the other i^+ gene, on the F factor, produces a normal repressor protein which is diffusible and capable of interacting with the operon to repress transcription. (See F19.2 in this book)

$i^+o^cz^+$ / F' o^+ = **Constitutive** because there is a constitutive operator (o^c) next to a normal z gene. Remembering that the operator functions in *cis* and is not influenced by the repressor protein, constitutive synthesis of β-galactosidase will occur.

$i^so^+z^+$ = **Repressed** because the product of the i^s gene is *insensitive* to the inducer lactose and thus can not be inactivated. The repressor will continually interact with the operator and shut off transcription regardless of the presence or absence of lactose.

$i^so^+z^+$ /F' i^+ = **Repressed** because, as in the previous case, the product of the i^s gene is *insensitive* to the inducer lactose and thus can not be inactivated. The repressor will continually interact with the operator and shut off transcription regardless of the presence or absence of lactose. The fact that there is a normal i^+ gene is of no consequence because once a repressor from i^s binds to an operator, the presence of normal repressor molecules will make no difference.

7. Refer to **K/C-Fig.19.4** and to F19.2 to get a good understanding of the lactose system before starting.

$i^+o^+z^+$ = Because of the function of the active repressor from the i^+ gene, and no lactose to influence its function, there will be **No Enzyme Made.**

$i^+o^cz^-$ = There will be a **Nonfunctional Enzyme Made** because even though the constitutive operator is in *cis* with a z gene, the z gene is mutant. The lactose in the medium will have no influence because of the constitutive operator. The repressor can not bind to the mutant operator.

$i^-o^+z^-$ = There will be a **Nonfunctional Enzyme Made** because with i^- the system is constitutive but the z gene is mutant. The absence of lactose in the medium will have no influence because of the non-functional repressor. The mutant repressor can not bind to the operator.

$i^-o^+z^-$ = There will be a **Nonfunctional Enzyme Made** because with i^- the system is constitutive but the z gene is mutant. The lactose in the medium will have no influence because of the non-functional repressor. The mutant repressor can not bind to the operator.

$i^-o^+z^+$ /F' i^+ = There will be **No Enzyme Made** because in the absence of lactose, the repressor product of the i^+ gene will bind to the operator and inhibit transcription.

$i^+o^cz^+$ /F' o^+ = Because there is a constitutive operator in *cis* with a normal z gene, there will be **Functional Enzyme Made.** The lactose in the medium will have no influence because of the mutant operator.

$i^+o^+z^-$ /F' $i^+o^+z^+$ = Because there is lactose in the medium, the repressor protein will not bind to the operator and transcription will occur. The presence of a normal z gene allows a **Functional Enzyme to be Made**. The repressor protein is diffusable, working in *trans*.

$i^-o^+z^-$ /F' $i^+o^+z^+$ = Because there is no lactose in the medium, the repressor protein (from i^+) will repress the operators and there will be **No Enzyme Made**.

$i^so^+z^+$ / F' o^+ = With the product of i^s there is binding of the repressor to the operator and therefore **No Enzyme Made.** The lack of lactose in the medium is of no consequence because the mutant repressor is insensitive to lactose.

$i^+o^cz^-$ /F' o^+z^+ = The arrangement of the constitutive operator (o^c) with the mutant z gene will cause a **Nonfunctional Enzyme to be Made.** In addition, the normal repressor will be inactivated by the lactose and the unit residing in the F factor will produce cause a **Functional Enzyme to be Made.** See F19.2.

8. First notice that in the first row of data, the presence of tm in the medium causes the production of active enzyme from the wild type arrangement of genes. From this one would conclude that the system is *inducible*. To determine which gene is the structural gene, look for the IE function and see that it is related to *c*. Therefore *c* codes for the **structural gene.** Because when *b* is mutant, no enzyme is produced, *b* must be the **promoter**.

Notice that when genes *a* and *d* are mutant, constitutive synthesis occurs, therefore one must be the operator and the other gene codes for the repressor protein. To distinguish these functions, one must remember that the repressor operates as a diffusible substance and can be on the host chromosome or the F factor (functionimg in *trans*). However, the operator can only operate in *cis*. In addition, in *cis*, the constitutive operator is dominant to its wild type allele, while the mutant repressor is recessive to its wild type allele.

Notice that the mutant *a* gene is dominant to its wild type allele, whereas the mutant *d* allele is recessive (behaving as wild type in the first row). Therefore, the *a* locus is the **operator** and the *d* locus is the **repressor** gene.

9. The *cI* gene is responsible for repressing the genes which control the lytic cycle of λ phage. If the *cI* gene is mutant then the lysogenic cycle could not occur. Interestingly, there are temperature sensitive alleles of the *cI* gene which maintain the lysogenic state at one temperature, but cause the lytic cycle at an elevated temperature.

10. In order to understand this question, it is necessary that you understand the negative regulation of the lactose operon by the *lac* repressor as well as the positive control exerted by the CAP protein. Remember, if lactose is present, it inactivates the *lac* repressor. If glucose is present, it inhibits adenyl cyclase thereby reducing, through a lowering or cAMP levels, the positive action of CAP on the *lac* operon.

(a) With no lactose and no glucose, the operon is off because the *lac* repressor is bound to the operator and although CAP is bound to its binding site, it will not override the action of the repressor.

(b) With lactose added to the medium, the *lac* repressor is inactivated and the operon is transcribing the structural genes. With no glucose, the CAP is bound to its binding site, thus enhancing transcription.

(c) With no lactose present in the medium, the *lac* repressor is bound to the operator region, and since glucose inhibits adenyl cyclase, the CAP protein will not interact with its binding site. The operon is therefore "off."

(d) With lactose present, the *lac* repressor is inactivated, however since glucose is also present, CAP will not interact with its binding site. Under this condition transcription is severely diminished and the operon can be considered to be "off."

11. Because the deletion of the regulatory gene causes a loss of synthesis of the enzymes, the regulatory gene product can be viewed as one exerting *positive control*. When tis is present, no enzymes are made, therefore, tis must inactivate the positive regulatory protein. When tis is absent, the regulatory protein is free to exert its positive influence on transcription.   Mutations in the operator negate the positive action of the regulator.   Below is a model which illustrates these points.

Model derived from results presented in problem #11.

POSITIVE CONTROL

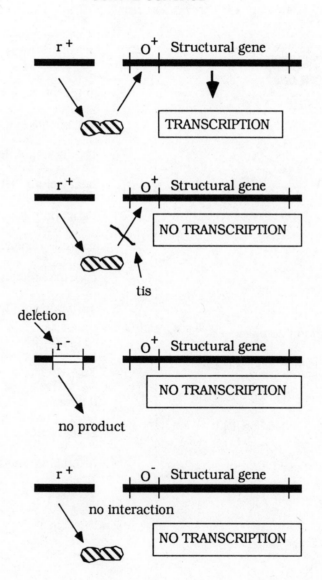

20

Genetic Regulation in Eukaryotes

Vocabulary: Organization and Listing of Terms

Structures and Substances

Operon

Introns

Chromatin

Inducer

DNA arrangements

    rRNA genes

    chorion genes

        nested replication complexes

    extended chromosome regions (ECRs)

    homogeneously staining regions (HSRs)

    double minute chromosomes (DM)

Antigen

Antibodies

B lymphocytes

T lymphocytes

T cell receptors

Immunoglobins

    IgG, IgA, IgM, IgD, IgE

    heavy chain (H)

    light chain (L)

    variable region

    constant region

    antibody combining sites

    hypervariable regions

    V genes

    J genes

    D segments

    recombination-activating gene (*RAG-1*)

    VDJ recombinase

Endosperm

Aleurone

Transposable controlling elements

    transposons, insertion sequences

        transposase

      rugosus

        starch-branching enzyme (SBEI)

    inverted repeats

    terminal sequences

    Alu family

Heterokaryons

    Sendai virus

    erythrocyte ghosts (EG)

Transcription factors

    functional domains

    Sp1

    zinc fingers

    homeodomain (helix-turn-helix)

      POU domain

      POU box

    leucine finger

    activator domains

RNA polymerase I, II, III

Promoters

Enhancers

Enhansons

Amphipathic alpha helix

    synthetic oligonucleotide (AH)

      scrambled order (SH)

    TFIID

Steroid hormones

    ecdysone

    receptor protein

    receptor-hormone complex

    hormone-responsive elements (HREs)

Transcripts

    α–amylase

    myosin

    α–crystallin

    immunoglobins

    preprotachykinin mRNA

    tachykinin

    α–, β–tubulins

      subunits

        colchicine

        vinblastine

      RNAse

      met-arg-glu-lys (MREI)

Oncogenes

Sarcoma

Retroviruses

    Rous sarcoma virus

    reverse transcriptase

    src

    acute transforming viruses

    nonacute or nondefective viruses

    v-onc, c-onc, proto-oncogene

    v-src, c-src

    v-ras

    v-mos

    v-erb

    c-abl

        chronic myelogenous leukemia (CML)

    c-myc

    c-jun, v-jun

    c-fos

Processes/Methods

Gene regulation levels

    organization of DNA

        amplification

            rRNA genes

            chorion genes

double minute chromosomes (DM)

    mechanisms

        DNA replication

            onionskin model

            nested replication complexes

        recombination and segregation

            sister chromatid exchange

            unequal exchange

gene rearrangement

    antibody variability
    (immune response)

        recombination of gene fragments

        recombination -activating
        gene *(RAG-1)*

        VDJ recombinase

transposable elements

    Dissociation (Ds)

    Activator (Ac)

        induced breakage

    rugosus

        insertion in exon

    hybrid dysgenesis

        P, I elements

    Alu family

transcription

    upstream (5') organization

promoters

    TATA box (-25 to -30)

    CAAT box (-70 to -80)

    GC box (-110)

       either orientation

    upstream regulatory sequences

enhancers

    variable position

    variable orientation

    cis-acting

    upstream and downstream

       upstream activator
       sequences (UAS)

transcription factors

    functional domains

       DNA binding domains

       structural motifs

         zinc fingers

         homeodomain (HD)

           (helix-turn-helix)

           POU domain

           POU box

         leucine zipper

    activator domains

positive regulation

    galactose metabolism

       kinase

       transferase

       epimerase

       regulatory loci

       UAS_9

    steroid hormones

       receptor-hormone complex

processing (post-transcriptional regulation)

    intron removal/exon splicing

    differential splicing (tissue specific)

transport

    mRNA stability

       autoregulation

         RNAse action

         ribosome stalling

translation

Cancer

    sarcoma

    retrovirus

    oncogenes

       conversion to proto-oncogenes

         point mutations

translocations

overexpression

    new promoters

    new upstream sequences

    amplification

DNA protection

Gel retardation

Affinity chromatography

Concepts

Genetic regulation

Antibody variability

    germline theory

    somatic mutation theory

    recombination theory

Stimulation of transcription

    relationships to transcription factors

Positive regulation

Autoregulation

Solutions to Problems and Discussion Questions

1. There are several reasons for anticipating a variety of different regulatory mechanisms in eukaryotes as compared to prokaryotes. Eukaryotic cells contain greater amounts of DNA and this DNA is associated with various proteins, including histones and nonhistone chromosomal proteins. *Chromatin* as such does not exist in prokaryotes. In addition, whereas there is usually only one chromosome in prokaryotes, eukaryotes have more than one chromosome all enclosed in a membrane (nuclear membrane). This nuclear membrane separates, both temporally and spatially, the processes of transcription and translation thus providing an opportunity for post-transcriptional, pre-translational regulation.

While prokaryotes respond genetically to changes in their external environment, cells of multicellular eukaryotes interact with each other as well as the external environment. The structural and functional diversity of cells of a multicellular eukaryote, coupled with the finding that all cells of an organism contain a complete complement of genes, suggests that in some cells certain genes are active which are not active in other cells.

It is often difficult to study eukaryotic gene regulation because of the complexities mentioned above, especially tissue specificity and the various levels at which regulation can occur (as indicated in question #2 below). Obtaining a homogeneous group of cells from a multicellular organism often requires a significant alteration of the natural environment of the cell. Thus, results from studies on isolated cells must be interpreted with caution. In addition, because of the variety of intracellular components (nuclear and cytoplasmic) it is difficult to isolate, free of contamination, certain molecular species. Even if such isolation is accomplished, it is difficult to interpret the actual behavior of such molecules in an artificial environment.

2. *Organization of DNA:* Changes in DNA/chromosome structure can influence overall gene output.

Gene amplification refers to cases where an increase in gene products is achieved by an increase in the number of genes producing those products. Such amplification can be achieved intrachromosomally (chorion genes) or extrachromosomally (rRNA genes in some amphibians).

Gene rearrangement involves the recombination of gene segments which changes the types and amounts of gene products (antibody variability).

Transposable elements may alter gene output by influencing promoters or enhancers or by introducing changes which alter gene coding or termination signals. It is also possible that transposable elements introduce novel promoters and/or enhancers.

Transcription: There are several factors which are known to influence transcription: *promoters* (TATA, CAAT, and GC boxes, as well as other upstream regulatory sequences; *enhancers*, which are *cis*-acting sequences which can act at various locations and orientations; *transcription factors*, with various structural motifs (zinc fingers, homeodomains, and leucine zippers) which bind DNA and influence transcription; *receptor-hormone complexes* which influence transcription.

Processing and transport types of regulation involve the efficiency of hnRNA maturation as related to capping, polyA tail addition, and intron removal.

Translation: After mRNAs are produced from the processing of hnRNA, they have the potential of being translated. The stability of the mRNAs appears to be an additional regulatory control point. Certain factors, such as protein subunits may influence a variety of steps in the translational mechanism. For instance, a protein or protein subunit may activate an RNAse which will degrade certain mRNAs or a particular regulatory element may cause a ribosome to stall, thus decreasing the speed of translation and increasing the exposure of a mRNA to the action of RNAses.

3. Evidence of cytoplasmic control of nuclear activity (or inactivity) has come from producing variations in nuclear-cytoplasmic relationships through the production of heterokaryons. Nuclei of a heterokaryon respond to cytoplasmic changes to such a degree that the reactivation of both DNA and RNA synthesis can be accomplished by changing the cytoplasmic environment. In addition, cellular morphology can change in response to cytoplasmic changes. Such cytoplasmic influences are not species specific.

4. *Promoters* are conserved DNA sequences which influence transcription from the "upstream" side (5') of mRNA coding genes. They are usually fixed in position and within 100 base pairs of the initiation site for mRNA synthesis. Examples of such promoters are the following: TATA, CAAT, and GC boxes.

Enhancers are *cis*-acting sequences of DNA which stimulate the transcription from most, if not all, promoters. They are somewhat different from promoters in that the position of the enhancer need not be fixed; it may be upstream, downstream, or within the gene being regulated. The orientation may be inverted without significantly influencing its action. Enhancers can work on different genes, that is, they are not gene-specific.

5. Transcription factors are proteins which are *necessary* for the initiation of transcription. However, they are not *sufficient* for the initiation of transcription. To be activated, RNA polymerase II requires four or five transcription factors. Transcription factors contain at least two functional domains: one binds to the DNA sequences of promoters and/or enhancers, the other interacts with RNA polymerase or other transcription factors.

6. The work of Cleveland and colleagues allowed selective changes to be made in the *met-arg-glu-lys* sequence. Only the engineered mRNA sequences which caused an amino acid substitution negated the autoregulation, indicating that it is the sequence of the amino acids, not the mRNA which is critical in the process of autoregulation.

Notice that code degeneracy allows for changes in mRNA sequence without changes in the amino acid sequence. Therefore, the model which depicts binding of factors to the nascent polypeptide chain is supported. There are a variety of experiments which could be used to substantiate such a model. One might stabilize the proposed MREI-protein complex with "crosslinkers," treat with RNAse to digest mRNA and to break up polysomes, then isolate individual ribosomes. One may use some specific antibody or other method to determine whether tubulin subunits contaminate the ribosome population.

7. Because the reading frame is shifted and the *met-arg-glu-lys* sequence is destroyed, one would expect that the autoregulation phenomenon would not occur. Without the appropriate binding sequence, the sequence of steps proposed in the model can not take place.

8. The cellular version of an oncogene is called a proto-oncogene. Comparison of oncogenes with proto-oncogenes reveals a variety of differences ranging from point mutations (*ras, mos*) to cases where portions of the proto-oncogene have been replaced or lost (*src, erb*). Not all oncogenes are different from the cellular genes.

9. Because immunoglobin genes have upstream regulatory sequences which enhance transcription, it is possible that certain translocations place in oncogenes in positions which allow a response to these regulatory sequences.

Part Three: Sample Test Questions
(with detailed explanations of answers)

Question 1. In 1956, Lederberg and coworkers, noted that wild type *E. coli* used as donors in λ-mediated transduction could only transduce the *gal⁺* gene but not other loci. Explain the proposed mechanisms for this type of transduction using fully labeled diagrams.

Concepts:

recombination in bacteria

transduction

    specialized

Answer 1. The case being discussed here involves *specialized transduction* in which generally only one genetic marker is transferred. In the case of λ, the attachment site for integration into the host (bacterial) chromosome is next to the *gal* locus. As indicated in the drawing below, the proximity of the λ attachment site (*att*) allows the *gal* locus to be included in the *defective* phage (λdgal) when the "incorrect" loop is formed upon prophage excision.

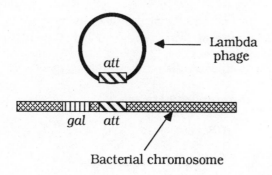

Common errors:

understanding of transduction

generation of λdgal

Question 2. Below are phrases which refer to various forms of recombination in bacteria. For each, clearly state whether you **agree** or **disagree**. If you disagree, briefly explain your reason(s).

(a) Transduction is the process in which exogenous DNA is drawn into bacteria as a single-stranded structure, then integrated into the bacterial chromosome.

(b) Generation of the F' state involves the production of a defective phage particle.

(c) Temperate phage are capable of entering a lysogenic cycle such that their genomes are incorporated into the bacterial chromosome.

(d) During the lysogenic cycle, phage are capable of producing bacteria when exposed to U.V. light.

Concepts:

recombination in bacteria

    transduciton

    transformation

    F' conjugation

    lysogeny

Answer 2.

(a) Disagree: The process being described refers to *transformation* not transduction. Transduction is *phage-mediated* recombination whereas in transformation exogenous DNA is taken up as indicated in the statement.

(b) Disagree: While generation of the F' state involves an insertion of a section of DNA into the bacterial chromosome (as does the production of a defective phage particle), the agent of insertion is an F factor. This process is *conjugational* not that related to lysogeny as is the production of a defective phage particle.

(c) Agree: Viruses which can enter either the lytic or lysogenic cycle are called *temperate* viruses. During the process of lysogeny, the viral chromosome is integrated into the bacterial chromosome as stated.

(d) Disagree: This statement is fairly silly in that it states that phage are capable of producing bacteria. Regardless of the exposure to U.V. light, phage can not produce bacteria. Ultraviolet light can cause induction of the lytic cycle, therefore phage, when lysogenic bacteria are exposed.

Common errors:

   carelessness in reading statements

   confusion as to what terms mean:

      transduction

      transformation

      conjugation

      lysogeny

      temperate viruses

Question 3. DNase is often used to map the locations where DNA-binding proteins (histones, RNA polymerases, transcription factors, etc.) interact with DNA. Restriction endonucleases are used to cut DNAs for identification and cloning. Why are these two different enzyme classes used in these different ways?

Concepts:

   experimental strategies

   action of nucleases

      DNase

   restriction endonucleases

Answer 3. DNase is a general term which includes a variety of exo- and endonucleases which cleave DNA from the ends or internally, respectively. Such cleavage is often irrespective of base sequence. If naked DNA is exposed to DNases, it is rapidly degraded to oligo- and mononucleotides. When protein is associated with DNA it protects regions from degradation. Such protected regions can be analyzed as to base content. Restriction endonucleases cleave DNA at specific sequences often hundreds or thousands or base pairs apart. If one is interested in mapping protein binding sites, one would want to use an enzyme (a DNase) with frequent yet relatively random cleavage characteristics, not restriction endonucleases.

Common errors:

   understanding overall strategy

   differences between

      DNases

   restriction endonucleases

Question 4. Assume that you have a cDNA clone for the gene causing retinoblastoma, and you prepare Southern blots probing DNA in cells from normal individuals and from children with retinoblastoma. Genomic DNA is prepared using the restriction endonuclease *Hind*III (the *Rb* gene contains four *Hind*III fragments as indicated below), and the following hybridization appears:

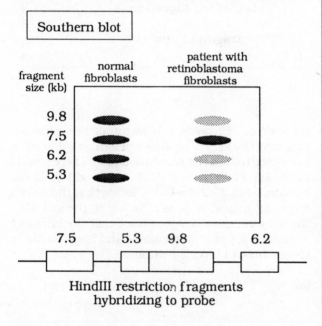

HindIII restriction fragments hybridizing to probe

What does each band represent?

What conclusions can be drawn from these data?

Concepts:

   experimental strategies

      cDNA probes

      Southern blots

      electrophoresis

      restriction endonuclease analysis

      hybridization

Answer 4. Notice that the fragment sizes (in kb) on the left side of the figure match the *Hind*III restriction fragments which are hybridizing (cDNA probe + genomic fragment) to the radioactive probe. Each band therefore represents a region where the radioactive probe is "trapped" by complementary base pairing to single-stranded DNA fragments which are bound to the filter. The smaller fragments migrate faster in the gel and therefore are in the bottom portion while the larger fragments are at the top, near the origin.

Notice that the intensity of the bands from the normal individual is somewhat uniform, indicating that all the restriction fragments are found in equal amounts. However, the intensity of three of the bands from the patient with retinoblastoma are about half as dense as in the normal. One band (7.5 kb) has the same intensity as in the normal.

Because humans are diploid organisms, with normally two copies of each gene, one may hypothesize that the individual with retinoblastoma has a heterozygous deletion of a portion of the retinoblastoma gene which includes *Hind*III fragments (9.8, 6.2, and 5.3 kb). It is likely that the inheritance of such a deletion is instrumental in causing familial retinoblastoma.

Common errors:

      understanding of experimental design

         cDNA probes

         Southern blots

         electrophoresis

         restriction endonuclease analysis

         hybridization

      recognition of deletion

Question 5. The *thioredoxin* gene in bacteria aids in the necessary reduction of proteins in *E. coli.* It encodes a protein of 108 amino acids and is contained in a 0.9 kb (*PstI/BamHI*) fragment. The gene for kanamycin resistance is contained in a 1.4 kb (*BamHI/ PstI*) fragment. The restriction map (one orientation) of these two genes (flanked by *BamHI* sites) in a plasmid vector is presented below.

(A)

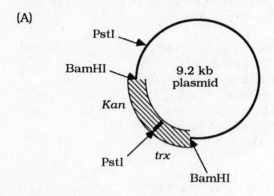

(a) Assume that the plasmid is restricted with the enzyme *BamHI.* What would be the electrophoretic pattern of the cleaved fragments?

(b) Assume that the orientation given above is only a guess and that the *Bam* HI fragment containing the *trx* and *Kan* genes could possibly exist in the opposite orientation (B). What experiment would you perform to determine whether the orientation is as in (A) or (B)?

(B)

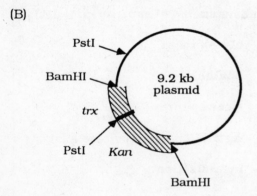

Concepts:

> **experimental strategies**
>
> **electrophoresis**
>
> **restriction digestion**
>
> **fragment analysis**

Answer 5. Below is a drawing of the expected product if either of the above plasmids (A) or (B) is restricted to completion with *BamHI.* There should be a 2.3 kb fragment (1.4 + 0.9 kb) and the remainder (9.2 - 2.3 = 6.9). Notice that the 6.9 kb fragment migrates slower (higher in the gel) than the 2.3 kb fragment. Also notice that the intensity of stain is less in the smaller band because there is less DNA to bind the stain.

(a)

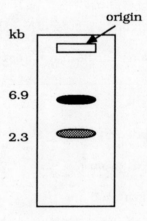

(b) To distinguish between the (A) and (B) orientations, one could make use of the change in position of the *PstI* restriction site in the two orientations. First, estimate the number of kb in the two fragments resulting from *PstI* restriction of orientation (A). Notice that in orientation (A) the *PstI* fragments are approximately 2.4 (1.4 for the *Kan* gene + about 1.0) kb and 6.8 kb. In orientation (B) the sizes would be approximately 1.9 (0.9 for the *trx* gene + about 1.0) and 7.3 kb. With appropriate standards, these size differences could be distinguished on agarose gels.

Common errors:

> **electrophoretic analysis**
>
> **restriction enzyme analysis**
>
> **experimental design**

Question 6. The Maxam and Gilbert DNA sequencing procedure involves chemical reactions (methylation) and cleavages (piperidine) that produce ^{32}P-labeled DNA fragments. The chemical reagents can give rise to G, G+A, C, and C+T cleavages. These fragments are then separated on electrophoretic gels, and the sequence is read directly from the gel. A sample gel is given below. From this gel, provide the sequence of the DNA fragment.

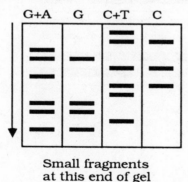

G+A   G   C+T   C

Small fragments
at this end of gel

Concepts:

> **electrophoresis**
>
> **DNA sequencing**

Answer 6. It is a relatively simple procedure to determine the sequence of DNA from the gel given. Start at the bottom with the smallest fragments. As one reads up the gel, one is reading from the 5' to the 3' direction. In the first case, note that there is a band in both the G+A and G lanes. Read this as a "G" because if an "A" occurs, it would exist as a band *only* in the G+A lane. The sequence would be as follows: 5'-GTGGTCACGACT

Common errors:

> **reading the gel as the sequence**
>
> **difficulty in dealing with G+A and C+T lanes**
>
> **careless mistakes**

Question 7. On the graph below draw C_0t curves for DNA from two genomes, one lacking repetitive DNA and the other containing repetitive DNA. Indicate the point on each curve at which the renaturation is half-complete. Label the horizontal and vertical axes accordingly. Explain the molecular basis for the different curves.

Answer 7. As discussed in the K/C text, the rate of reassociation of melted DNA increases as the proportion of repetitive DNA increases.   Such relationships are reflected in $C_o t$ curves in which one plots the fraction of DNA reassociated against the a logarithmic scale of $C_o t$ values which have the units (mole X sec/liter). Because denatured DNA strands which are repetitive have a higher likelihood of complementary interaction, the time of reassociation is less.

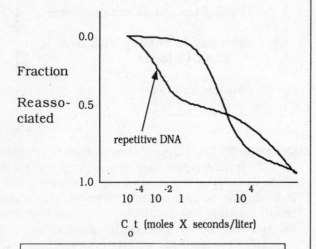

$C_o t$ (moles X seconds/liter)

Question 8. Below is a series of boxes containing a hypothetical gene depicting two subloci (cistrons) which each produce different products that unite to form a functional (wild type) protein. An "o" indicates the site of a mutation. In the space to the right of each box, indicate whether a *wild type* or *mutant* phenotype is expected.

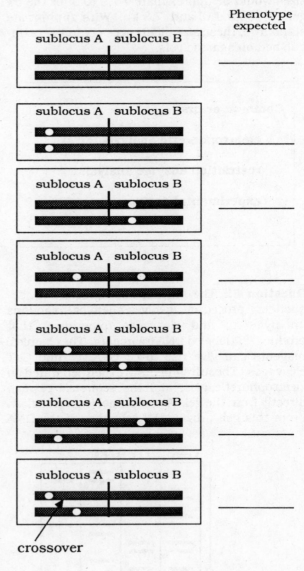

Phenotype expected

crossover

> **Concepts:**
>
> > **gene concepts**
> >
> > **cistron**
> >
> > **complementation**
> >
> > **intragenic crossing over**
> >
> > **pseudoalleles**

Answer 8. This problem deals with two concepts relating to gene structure: (1) complementation and (2) intragenic crossing over. Two genes (cistrons) are said to complement each other if in the "diploid" state where each chromosome contains a mutant allele for the gene in question, a wild type function is observed.

In the problem, it is stated that the two subloci produce products which unite to form the wild type function. Therefore, in all of the cases where there is *at least* one wild type "A" sublocus and one wild type "B" sublocus, a wild type function will result. In the last case, a crossover in the position of the arrow will generate two "A" subloci, one with both mutations and one with no mutations. In this case therefore, a wild type product is possible.

List of answers:

> wild
> mutant
> mutant
> wild
> wild
> wild
> *rare* wild

> **Common errors:**
>
> > **concept of complementation**
> >
> > **intragenic crossing over**

Question 9. Depending on the regulatory system, in prokaryotes when the regulatory protein **is** or **isn't** bound to the DNA, the operon may be **on** or **off**. Fill in the chart below and give a brief explanation of your reasoning.

Relationship of Regulator Protein to DNA	*Operator*	
	Positive control	*Negative control*
is bound		
isn't bound		

> **Concepts:**
>
> > **genetic regulation in prokaryotes**
> >
> > **positive vs. negative control**

Answer 9. Refer to F19.1 in this book and notice that any time a regulatory protein interacts with DNA and transcription is stimulated, it is called *positive* control. Any time a regulatory protein interacts with DNA and represses transcription it is called *negative* control. In completing the chart, apply these simple rules.

Relationship of Regulator Protein to DNA	Operator	
	Positive control	*Negative control*
is bound	**on**	**off**
isn't bound	**off**	**on**

Common errors:

    confusion with positive and negative control

    confusion with repressible and inducible systems

Question 10. Acquired Immune Deficiency Syndrome (AIDS) is caused by a retrovirus which exerts its pathogenesis by immunosuppression and selective depletion of helper T lymphocytes. One strategy for AIDS therapy is administration of AZT (azidothymidine) which blocks reverse transcriptase. Present the "life cycle" of a typical retrovirus and show how AZT may act as a therapeutic agent.

Concepts:

    retroviral life cycle

    action of reverse transcriptase

Answer 10. Retroviruses contain an RNA genome, which in the "life" cycle, is "reverse transcribed" into a DNA strand. First, the single-stranded RNA genome, in the presence of reverse transcriptase, forms a complementary DNA strand. From the DNA strand, a complementary DNA strand is made thus giving a DNA double-stranded structure which can be integrated into the host DNA. AZT is an inhibitor of reverse transcriptase and therefore blocks a critical step in the cycle mentioned above. Without activity of reverse transcriptase, the RNA genome can not manufacture the complementary strand of DNA.

Common errors:

    If students understand the "life" cycle of the virus (retrovirus) there are few difficulties with this type of problem.

21

Developmental Genetics

Vocabulary: organization and listing of terms

Structures and Substances

Receptors

Polytene chromosomes

    chromosome puffs

Isozymes

    lactate dehydrogenase

    tetrameric

Growth hormone

    pituitary gland (anterior)

    GHF-1 transcription factor

    homeodomain

    POU domain

Nucleosome

Cyclic AMP

Chemotactic gradients

    differentiation-inducing factor (DIF)

Molecular gradients

    anterior-posterior axis

        bicoid mRNA

        positive transcriptional regulator

    maternal genes

Maternal cytoplasm

Cortex

Blastoderm

Imaginal disks

Blastemal mesenchyme

Processes/Methods

Development

    determination

        regulatory events

        patterns of gene activity

differentiation

    genetic and morphological changes

cell-cell interaction

    intercellular communication

    cell migration

    gradients and patterns

cellular specialization

    nuclear differentiation

    tissue-specific activation

morphogenesis

extracellular environment

    maternal cytoplasm

Analyses

biochemical and cytophotometric

newt embryo (Spemann)

Rana pipiens (Briggs and King)

Xenopus laevis (Gurdon)

    enucleation of oocytes

    serial transfers

    gene reactivation

nonmeristematic cells

    callus

Chironomus

    polytene chromosomes

    puff patterns

    differential transcription

isozymes

    lactate dehydrogenase

growth hormone

    in situ hybridization

    immunochemistry

bacterial sporulation

    mutation analysis

    RNA polymerase

        sigma factors

Dictyostelium discoidium

    myxamoebae

    slug

    chemotactic gradients

    aggregation center

heterokaryon

Drosophila

    embryogenesis

    blastoderm

    imaginal disks

    fate maps

compartments

metamorphosis

molecular gradients

anterior-posterior

dorsal-ventral

selector genes

homeotic mutants

positional cues

molecular cloning

(recombinant DNA techniques)

in situ hybridization

Concepts

Development (F21.1)

determination

differentiation

cell-cell interaction

Developmental genetics:

molecular explanations

presence or absence of molecules

receptors

transcriptional events

different cells

different times

cell and tissue interactions

observable morphological events

Heuristic concepts

Variable gene activity (F21.2)

Stability of differentiation

transdetermination

regeneration

modulation

Genomic Equivalence

Totipotent

genomic repression

reprogrammed genes

nuclear differentiation

Maternal influences

anterior-posterior gradient

positional information

Cytoplasmic influences

heterokaryon

Homeotic genes

F21.1. Illustration of the relationship between *determination* and *differentiation*. Determination sets the program which will later be revealed during differentiation.

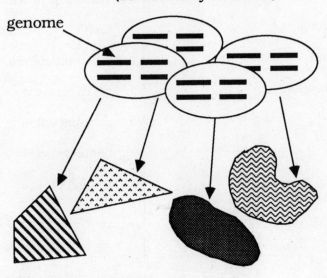

Cells become DETERMINED
(structurally uniform)

genome

Cells DIFFERENTIATE
(structurally different)

F21.2. Illustration of the genomic changes which are thought to occur during cell differentiation. The *variable gene activity model* states that different sets of genes are transcriptionally active in differentiated cells.

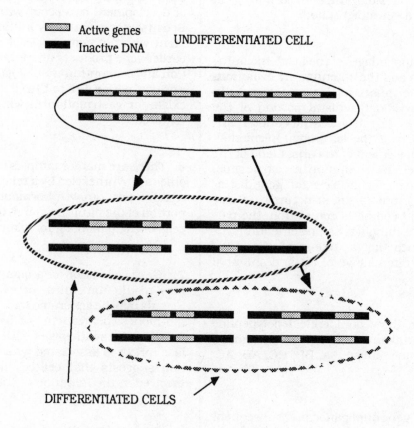

Solutions to Problems and Discussion Questions

1. *Determination* refers to early developmental and regulatory events which set eventual patterns of gene activity. Determination is not the end result of the regulatory activity, rather, it is the process by which the genomic fate of a particular cell type is fixed. *Differentiation* on the other hand follows determination and is the manifestation, in terms of genetic, physiological, and morphological changes, of the determined state.

2. Many of the appendages of the head, including the mouth parts and the antennae are evolutionary derivatives of ancestral leg structures. In *spineless aristapedia* the distal portion of the antenna is replaced by its ancestral counterpart, the distal portion of the leg (tarsal segments). Because the replacement of the arista (end of the antenna) can occur by a mutation in a single gene, one would consider that one "selector" gene distinguishes aristal from tarsal structures. Some support for this hypothesis comes from the pattern of transdetermination as presented in K/C-Fig.21.20. Notice that a "one-step" change is involved in the interchange of leg and antennal structures.

3. If there are three nonidentical polypeptide chain subunits, taken two at a time there can be 6 different isozyme forms: AA, BB, CC, AB, AC, BC.

4. The theory of gene duplication and subsequent evolution referred to in an earlier chapter could apply to the concept of isozymes because they are functionally related molecules which are electrophoretically distinct. It would be efficient for functionally related molecules to develop by gene duplication because the original function can be accomplished while the duplicated gene is altered, perhaps in stages, by mutation.

To test for molecular homology, one could determine amino acid composition and sequence. Isozymes originated by gene duplication should show similarity in these two aspects.

5. Actinomycin D is useful in determining the involvement of transcription on molecular and developmental processes. Because maternal RNAs are present in the fertilized sea urchin egg (as is the case with many egg types) a considerable amount of development can occur without transcription. Because gastrulation is inhibited by *prior* treatment with actinomycin D, it would appear that earlier gene products are necessary for the initiation and/or continuation of gastrulation. Clearly a critical period (6th to 11th hours of development) exists for gastrulation in which gene activity is required.

6. There are many examples where certain molecules are synthesized by a relatively small number of cell types: growth hormone from the anterior portion of the pituitary gland, insulin from certain cells of the pancreas, for examples.

7. Gene activation is a complex process which may result from intra- or extracellular signals. Signals may be generated form other, distant cells (hormones, nerves, etc.) or by cell contact. The fact that the synthesis of UDPG-pyrophosphorylase mRNA is associated with cellular reaggregation suggests that cell-cell interaction may be involved in the initiation of transcription.

8. The fact that nuclei from almost any source remain transcriptionally and translationally active substantiates the fact that the genetic code and the ancillary processes of transcription and translation are compatible throughout the animal kingdom.

Because the egg represents an isolated, "closed" system which can be mechanically, environmentally, and to some extent biochemically manipulated, various conditions may be developed which allow one to study facets of gene regulation. For instance, the influence of transcriptional enhancers and suppressors may be studied along with factors which impact on translational and post-translational processes. Combinations of injected nuclei may reveal nuclear-nuclear interactions which could not normally be studied by other methods.

9. The egg is not an unorganized collection of molecules from which life springs after fertilization. It is a highly organized structure, "preformed" in the sense that maternal informational molecules are oriented to provide an anterior-posterior and dorsal-ventral pattern from which nuclei receive positional cues.

Such positional cues lead to the "determined" state, from which cells later reveal their adult form (differentiation). Indeed, in *Drosophila* and many other organisms, embryonic fate maps may be constructed thereby attesting to the maternally-derived "prepattern" present in the egg.

The egg therefore *is* preformed, not in the sense that a miniature individual resides, but in a molecular prepattern upon which development depends. However, work by Spemann, Briggs and King, and Gurdon, indicates that there is plasticity in the programming of nuclei and that even nuclei from somewhat specialized cells often have the potential to direct the development of the entire adult individual. Such *totipotent* behavior of cells indicates that development arises as a result of a series of progressive steps in which cells acquire new structures and functions as development progresses. The epigenetic theory is viewed today as the result of differential gene expression.

22

Extrachromosomal Inheritance

Vocabulary: Organization and Listing of Terms

Structures and Substances

Kynurenine

Tryptophan

Heterokaryon

Hyphae

Mycelia

Conidia

Paramecin

Kappa

Processes/Methods

Extrachromosomal (Extranuclear) influence

    maternal effect

        Ephestia kuehniella (A, a)

        Limnaea peregra (D, d)

           dextral, sinistral

first cleavage division

spindle orientation

injection experiments

Drosophila melanogaster (*fu*, *fu*+)

organelle heredity (maternal inheritance)

chloroplast DNA

    Mirabilis jalapa

    Maize (*iojap*)

    Chlamydomonas reinhardi (*sr*)

mitochondrial DNA

    Neurospora crassa (*poky*)

        suppressive

    Saccharomyces cerevisiae (*petite*)

        segregational

        neutral

        suppressive

Humans

    heteroplasmy

    myoclonic epilepsy (MERRF)

    Leber's hereditary optic

        neuropathy (LHON)

    Kearns-Sayre syndrome

infectious heredity

    Paramecium aurelia

        killers

        paramecin

        kappa

        conjugation

        autogamy

Drosophila

    CO_2 sensitivity

D. bifasciata, D. willistoni

    sex ratio

Concepts

Extrachromosomal influence (F22.1)

Non-Mendelian patterns
of inheritance (F22.1)

F22.1. Illustration of the common pattern seen in many cases of extrachromosomal inheritance. The condition of the egg parent has a stronger influence on the phenotype of the offspring than the sperm/pollen parent. Therefore reciprocal crosses give different results.

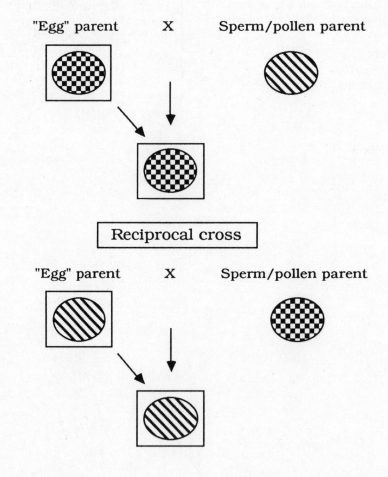

Solutions to Problems and Discussion Questions

1. In cases of extrachromosomal inheritance, the phenotype is determined by the genetic (maternal effect) or cytoplasmic (organelle or infectious) condition of the parent which contributes the bulk of the cytoplasm to the offspring. In most cases, the maternal parent provides the basis for the cytoplasmic inheritance.

The pattern of inheritance is more often from one parent to the offspring. One does not see both parents contributing to the characteristics of the offspring as is the case with Mendelian (chromosomal) forms of inheritance. Standard Mendelian ratios (3:1) are usually not present. In general, the results of reciprocal crosses differ. See F22.1.

Female mutant   X   male wild

all offspring mutant

Female wild   X   male mutant

all offspring wild

In sex-linked inheritance, the pattern is often from grandfather through carrier mother to son. Patterns of extrachromosomal inheritance are often not influenced by the sex of the individual.

2. The case with *Limnaea* involves a maternal effect in which the *genotype* of the mother influences the *phenotype* of the *immediate* offspring in a non-Mendelian manner. Notice that in the above statement, it is the maternal genotype which determines the phenoytpe of the offspring, regardless of its own genotype.

Since both of the parents are *Dd*, the parent contributing the eggs must be *Dd*. Therefore, all of the offspring must have the phenotype of the mother's genotype, which is dextral.

3. The *mt*⁺ strain (resistant for the nuclear and chloroplast genes) contributes the "cytoplasmic" component of streptomycin resistance which would negate any contribution from the *mt*⁻ strain. Therefore, all the offspring will have the streptomycin resistance pheotype. In the reciprocal cross, with the *mt*⁺ strain being streptomycin sensitive, the nuclear genome of half of the offspring will contain a streptomycin resistance gene and therefore be resistant.

4. Because the ovule source furnishes the cytoplasm to the embryo and thus the chloroplasts, the offspring will have the same phenotype as the plant providing the ovule.

a) green

b) white

c) variegated (patches of white and green)

d) green

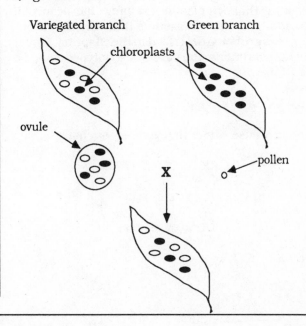

5.   As with any description of dominance, one looks to the phenotype of the diploid heterozygote. In this problem, the heterozygote is of normal phenotype, therefore the *petite* gene is recessive.

6.   Examine K/C-Fig.22.5 and notice that the inheritance patterns for the two, *segregational* and *neutral*, are quite different. The segregational mode is dependent on nuclear genes while that of the neutral type is dependent on cytoplasmic influences, namely mitochondria. If the two are crossed as stated in the problem, then one would expect, in the diploid zygote, the *segregational* allele to be "covered" by normal alleles from the neutral strain.

On the other hand, as the nuclear genes are again "exposed" in the haploid state of the ascospores, one would expect a 1:1 ratio of normals to petites. The petite phenoytpe is caused by the nuclear, *segregational* gene.

7.   In providing the answers to this question, remember that a *Paramecium* may be sensitive and carry the *K* gene. These are organisms which did not obtain kappa particles. There is a question as to whether cytoplasmic exchange has occurred, however, seeing the results in cross (b) one can assume that cytoplasmic exchange has occurred. In addition, one can assume that only the exconjugants are being described in the offspring. Under those conditions, the parental genotypes could be the following:

(a) *Kk X kk*

(b) any case where this is no *kk* such as:

KK   X   KK

or

KK   X   kk

(c) *Kk X Kk*

8. **(a)** There are many similarities among mitochondrial, chloroplast, and prokaryotic molecular systems. It is likely that mitochondria and chloroplasts evolved from bacteria in a symbiotic relationship, therefore it is not surprising that certain antibiotics which influence bacteria will also influence all mitochondria and chloroplasts.

(b) Clearly, the mt^+ strain is the donor of the cpDNA since the inheritance of resistance or sensitivity is dependent on the status of the mt^+ gene.

9. In a maternal effect, the *genotype* of the mother influences the *phenotype* of her immediate offspring in a non-Mendelian manner. The fact that all of the offspring (F_1) showed a dextral coiling pattern indicates that one of the parents (maternal parent) contains the *D* allele. Taking these offspring and seeing that their progeny (call these F_2) occur in a 1:1 ratio indicates that half of the offspring (F_1) are *dd*. In order to have these results, one of the original parents must have been *Dd* while the other must have been *dd*.

Parents: *Dd X dd*	
Offspring (F_1): 1/2 *Dd*, 1/2 *dd*	
(all dextral because of the maternal genotype)	
Progeny (F_2):	
All those from *Dd* parents will be dextral while all those from *dd* parents will be sinistral.	

10. It appears as if some factor normally provided by the gs^+ allele is necessary for normal development and/or functioning of the female offspring's gonads. Without this product, the daughters are sterile, thus the term "grandchildless." Because the female provides so much vital material and information to the egg, including the cytoplasm necessary for germ line determination, it is not surprising that such maternal effect genes exist.

23

Quantitative Genetics

Vocabulary: Organization and Listing of Terms

Structures and Substances

Lactose

DDT

Processes/Methods

Discontinuous traits

Continuous traits

    biometry

    multiple-factor hypothesis

    additive

    polygenic

        phenotypic flexibility

        interaction with the environment

            resistance to DDT (*Drosophila*)

Statistical analysis

    descriptive summary

statistical inference

statistics

parameters

    mean

        central tendency

    variance

    standard deviation

    standard error of the mean

Phenotypic expression

    penetrance

    expressivity

    genetic background

        genetic suppression

        position effects

            translocation

            heterochromatin

temperature effects

  conditional

    temperature sensitive

      lethal

      restrictive temperature

      permissive temperature

nutritional mutants

  phenylketonuria

  galactosemia

  diabetes

  hypoglycemia

  lactose intolerance

onset

  Tay-Sachs disease

  Lesch-Nyhan syndrome

  Huntington's disease

Heredity vs. environment

  broad heritability

  narrow heritability

    monozygotic (identical) twins

    dizygotic (fraternal) twins

      concordant

      discordant

Concepts

Heritability

Phenotypic variation

Heredity vs. environment

  inbred strains

  isogenic

  heritability ratio (H^2)

    broad heritability

      environmental variance

      genetic variance

      interaction

    narrow heritability

      dominant genes

      additive genes

Solutions to Problems and Discussion Questions

1. In *polygenic* inheritance two or more gene pairs are *combining* their influence in an additive manner to produce the phenotype. With *epistasis*, one gene pair acts to *mask* the expression of another gene pair. In *discontinuous* variation the influences of each gene pair are not additive and more typical Mendelian ratios such as 9:3:3:1 and 3:1 result. In *continuous* variation, different gene pairs interact (uaually additively) to produce a phenotype which is less "stepwise" in distribution.

2. Many traits, especially those which we view as quantitative are likely to be determined by a polygenic mode. The following are some common examples: height, general body structure, skin color, and perhaps most common behavioral traits including intelligence.

3. *Penetrance* refers to the percentage of individuals which express the mutant genotype while *expressivity* refers to the range of expression of a given genotype.

4. (a,b) Both *position effect* and *suppressor genes* are conditions of the genetic background of an organism which influence the expression of a gene. Position effects may occur when a gene is moved (translocation or inversion) from one chromosomal location to another. As yet unknown factors, especially from neighboring heterochromatin, alter the expression of a "moved" gene.

While the molecular action of many suppressor genes is well understood in prokaryotes, little is known about their action in eukaryotes except that, when present they suppress the mutant action of non-allelic genes.

For instance, the *vermilion* gene in *Drosophila* causes a change in the eye color; however with *su-v* (*suppressor of vermilion*) present, the eye color is wild type. Thus the genetic background of an organism may have a considerable influence on the penetrance and expressivity of a gene.

(c) *Monozygotic twins* are derived from a single fertilized egg and are thus genetically identical to each other. They provide a method for determining the influence of genetics and environment on certain traits. *Dizygotic twins* arise from two eggs fertilized by two sperm cells. They have the same genetic relationship as siblings.

The role of genetics and the role of the environment can be studied by comparing the expression of traits in monozygotic and dizygotic twins. The higher concordance value for monozygotic twins as compared to the value for dizygotic twins indicates a significant genetic component for a given trait.

(d) *Concordance* refers to the frequency with which both members of a twin pair express a given trait. *Discordance* refers to the frequency at which one twin expresses a trait while the other does not. A comparison of concordance (and discordance) frequencies can provide information on the genetic and/or environmental influence on a given trait.

(e) *Heritability* is a measure of the degree to which the phenotypic variation of a given trait is due to genetic factors. A high heritability indicates that genetic factors are major contributers to phenotypic variation while environmental factors have little impact.

5. For height, notice that average differences between MZ twins reared together (1.7 cm) and those MZ twins reared apart (1.8 cm) are similar (meaning little environmental influence) and considerably less than differences of DZ twins (4.4 cm) or sibs (4.5) reared together. These data indicate that genetics plays a major role in determining height.

However, for weight, notice that MZ twins reared together have a much smaller (1.9 kg) difference than MZ twins reared apart indicating that the environment has a considerable impact on weight. By comparing the weight differences of MZ twins reared apart with DZ twins and sibs reared together one can conclude that the environment has almost as much an influence on weight as genetics.

6.

P:

$v^+/v^+; su\text{-}v/su\text{-}v$ X $v/Y; su\text{-}v^+/su\text{-}v^+$

F_1:

$v^+/v; su\text{-}v^+/su\text{-}v$ (females, wild type)

$v^+/Y; su\text{-}v^+/su\text{-}v$ (males, wild type)

F_2:

using the forked-line method

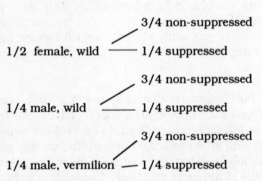

Multiplying through gives:

= 3/8 female wild

= 1/8 female wild

= 3/16 male wild

= 1/16 male wild

= 3/16 male vermilion

= 1/16 male wild

Overall, 1/2 of the offspring are wild type females, 5/16 are wild type males, and 3/16 are vermilion males.

7. At first glance, this problem looks as if it will be an arithmetic headache, however, the problem can be simplified .

(a) The mean is computed by adding the measurements of all of the individuals, then dividing by the number of individuals. In this case there are 760 corn plants. To keep from having to add 760 numbers, merely multiply each height group by the number of individuals in each group. Add all the products then divide by n (760). This give a value for the mean of 140 cm.

(b) For the variance, use the formula given below (as in the text):

$$s^2 = V = (\Sigma X_i^2 - n\overline{X}^2) / n - 1$$

To simplify the calculations, determine the square of each height group (100 cm for example) then multiply the value by the number in each group. For the first group (100 cm) we would have:

100 X 100 X 20 = 200000

The rest of the groups are as follows:

```
110 X 110 X 60  = 726000
120 X 120 X 90  = 1296000
130 X 130 X 130 = 2197000
140 X 140 X 180 = 3528000
150 X 150 X 120 = 2700000
160 X 160 X 70  = 1792000
170 X 170 X 50  = 1445000
180 X 180 X 40  = 1296000

                = 15180000
```

Now, the mean squared, then multiplied by n is as follows:

$$140 \text{ X } 140 \text{ X } 760 = 14896000$$

Completing the calculations gives the following:

$$(15180000 - 14896000)/759$$

$$= 284000/759$$

$$= 374.18$$

(c) The *standard deviation* is the square root of the variance or 19.34.

(d) The *standard error of the mean* is the standard deviation divided by the square root of n, or about 0.70.

8. If you add the numbers given for the ratio, you obtain the value of 16 which is indicative of a dihybrid cross. The distribution is that of a dihybrid cross with additive effects.

(a) Because a dihybrid result has been identified, there are two loci involved in the production of color. There are two alleles at each locus for a total of four alleles.

(b) Because of the 1:4:6:4:1 ratio, the effects are additive. Recall that if dominance occurs, a 9:3:3:1 ratio is obtained. Epistasis produces ratios like 9:3:4, or 9:7.

(c) Because the description of red, medium-red, etc. gives us no indication of a *quantity* of color in any form of units, we would not be able to actually quantify a unit amount for each change in color. We can say that each gene provides an equal *unit* amount to the phenotype and the colors differ from each other in multiples of that unit amount.

(d) The genotypes are as follows:

1/16	= dark red	=	AABB
4/16	= medium-dark red	=	2AABb 2AaBB
6/16	= medium red	=	AAbb 4AaBb aaBB
4/16	= light red	=	2aaBb 2Aabb
1/16	= white	=	aabb

9. (a) It *is possible* that two parents of moderate height can produce offspring that are much taller or shorter than either parent because segregation can produce a variety of gametes, therefore offspring as illustrated below:

rrSsTtuu X *RrSsTtUu*
(moderate) (moderate)

Offspring from this cross can range from very tall *RrSSTTUu* (12 "tall" units) to very short *rrssttuu* (8 "tall" units).

(b) If the individual with a minimum height, *rrssttuu*, is married to an individual of intermediate height *RrSsTtUu*, the offspring can be no taller than the height of the tallest parent. Notice that there is no way of having more than four dominant alleles in the offspring.

10. The formula for estimating heritability is

$$H^2 = V_G / V_P \text{ where } V_G \text{ and } V_P$$

are the genetic and phenotypic components of variation, respectively. The main issue in this question is obtaining some estimate of two components of phenotypic variation: genetic and environmental.

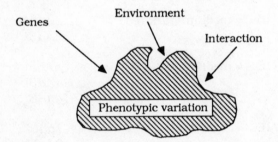

V_P is the combination of genetic and environmental variance. Because the two parental strains are true-breeding, they are assumed to be homozygous and the variance of 3.124 and 3.876 considered to be the result of environmental influences. The average of these two values is 3.5. The F_1 is also genetically homogeneous and gives us an additional estimation of the environmental factors. By averaging with the parents

$$[(3.500 + 4.743)/2 = 4.123]$$

we obtain a relatively good idea of environmental impact on the phenotype. The phenotypic variance in the F_2 is the sum of the genetic (V_G) and environmental (V_E) components. We have estimated the environmental input as 4.123, so 47.708 minus 4.123, gives us an estimate of (V_G) which is 43.585. Heritability then becomes 43.585/ 47.708 or 0.914. This value, when viewed in percentage form indicates that about 91% of the variation in corolla length is due to genetic influences.

24

Behavior Genetics and Neurogenetics

Vocabulary: Organization and Listing of Terms

Structures and Substances

Alcohol dehydrogenase

Acetaldehyde dehydrogenase

Transducers

Ion channels

    sodium

    potassium

Calmodulin

Ethylmethanesulfonate

Processes/Methods

Behavior

    innate (instinctive)

    previous environmental setting

        behaviorist school

    nature-nurture controversy

Methodologies

    comparative approach

        alcohol preference in mice

        open-field behavior in mice

            pleiotropic

        baboon mate selection

    selection

        maze learning in rats

        geotaxis in *Drosophila*

            taxes: positive or negative

        polygenic control

    effects of single genes

        nest-cleaning behavior in honeybees

            hygienic behavior

taxes in bacteria

   chemotaxis

      excitation pathway

      adaptation pathway

avoidance reactions in *Paramecium*

behavior genetics of a Nematode

   chemotaxis

   thermotaxis

   generalized movement

single-gene effects in mice

genetic dissection/mosaic analysis

   courtship

      orientation

      vibration

electroretinogram

   primary focus (F24.1)

      ring-X chromosome (F24.1)

   blastoderm

Learning in *Drosophila*

   memory-deficient mutant

Molecular biology of behavior

Human behavior genetics

   clear-cut genetic basis

   less-defined genetic bases

Concepts

Nature/nurture controversy

Primary focus of a gene (F24.1)

F24.1. Illustration of the procedure of using the ring-X chromosome and appropriate markers (*singed bristles*) to generate gyndromorphic flies for mapping the primary focus of a gene (*limp, lp*). After elimination of the unstable ring-X chromosome in one of the daughter cells at the first mitotic division, male (XO) and female (XX) tissues are produced in the same fly. Having the *singed bristles* gene in the heterozygous state, allows one to identify the male and female tissues. Male tissue will show the singed bristles phenotype.

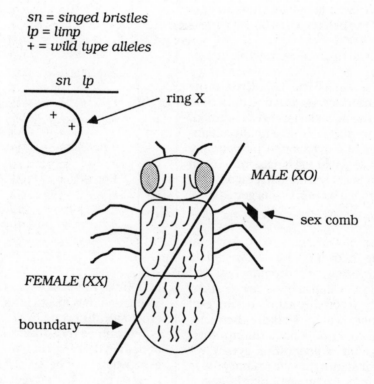

sn = singed bristles
lp = limp
+ = wild type alleles

Solutions to Problems and Discussion Questions

1. The first method described in the K/C text attempts to correlate behavioral differences with general genetic differences. It is a comparative approach in which a particular behavior is examined among several (to many) closely related but genetically different strains of organisms. If the environment is held constant, yet the behavioral differences persist, there must be a genetic component to the behavior.

The second approach involves *selection* for certain behaviors. When selection and interbreeding of a behavior yields a consistent phenotype, then the genetics of the behavior can be examined.

The third approach takes advantage of the fact that some behaviors are strongly influenced by a single (or few) locus. Under this condition, mutagens can be used to induce mutations from which those of interest to the behaviorist can be studied. While this method produces the genetically "cleanest" path of analysis, it provides information only on those behaviors which are strongly influenced by few gene pairs. It is likely that many significant behaviors are controlled by numerous gene pairs.

2. The advantages of *Drosophila* in behavioral studies include the following. First, *Drosophila* has an immense repertoire of chromosomal and single-gene alterations which allows the investigator the opportunity to create a broad range of genetic circumstances which facilitate the isolation and characterization of behavioral mutants. A good example is that of generating gynandromorphs (mosaics) for identifying the primary focus of a gene which determines a particular behavior (see F24.1). Such an analysis is not possible in *Caenorhabditis* at this time.

Second, *Drosophila* has a fairly elaborate set of various behaviors (reproductive, locomotor, taxic, etc.) which allows one to generate and isolate a variety of abnormal behaviors upon which genetic studies depend.

However, because of the relative developmental simplicity of *Caenorhabditis*, most scientists view this organisms as eventually contributing greatly to our understanding of such complex processes as development and behavior.

3. Examine the data and notice that the magnitude of change attributed to the X chromosome is relatively small. Notice that chromosome 2 contributes relatively strongly to negative geotaxis, while chromosome 3 contributes fairly strongly to positive geotaxis.

4. Refer to K/C-Fig.24.7 in answering this question. In a mating of *uR* X *UuRR*, the following offspring are produced:

1/2 *UuRR* = nonhygienic
1/2 *uuRR* = uncap only

For *Ur* X *UURr*:

1/2 *UURr* = nonhygienic
1/2 *UUrr* = remove only

For *uR* X *Uurr*:

1/2 *UuRr* = nonhygienic
1/2 *uuRr* = uncap only

5. One of the easiest ways to determine whether a genetic basis exists for a given abnormality is to cross the abnormal fly to a normal fly. If the trait is determined by a dominant gene, the trait should appear in the offspring, probably half of them if the gene was in the heterozygous state. If the gene is recessive and homozygous, then one may not see expression in the offspring of the first cross, however if one crosses the F_1 one should see the trait appearing in approximately 1/4 of the offspring. Modifications of these patterns would be expected if the mode of inheritance is X-linked or shows other modifications of typical Mendelian ratios.

One might hypothesize that the focus is in the nervous or muscular system. Mapping the primary focus of the of the gene could be accomplished with patience and the use of the unstable ring-X chromosome to generate gynandromorphs. Given that the gene is X-linked, one would use classical recombination methods to place a recessive X-linked marker, such as *singed bristles*, on the X chromosome with the gene causing the limp. This would help one identify the male/female boundaries. One would then cross homozygous (for the trait and markers) females to ring-X males, or the reciprocal, then examine the offspring for gynandromorphs (and singed mosaics). If one obtained a pool of gynandromorphs, one could then assess the phenotype (limp or normal) with respect to exposure of the recessive gene in the male tissue. Correlating such would allow one to provide an educated guess as to the primary focus of the gene causing the limp. See F24.1.

6. From the information provided, one can conclude that the tasting trait is determined by a dominant gene. Notice that a 3:1 ratio of tasters to nontasters is obtained in the third cross. This result strongly argues for the *TT* or *Tt* condition providing taste of PTC. Information provided by the other crosses reinforces this hypothesis.

7. Several problems in the study of human behavioral genetics would include the following.

1. With a relatively small number of offspring produced per mating, standard genetic methods of analysis are difficult.

2. Records on family illnesses, especially behavioral illnesses, are difficult to obtain.

3. The long generation time makes longitudinal (transmission genetics) studies difficult.

4. The scientist can not direct matings which will provide the most informative results.

5. The scientists can't always subject humans to the same types of experimental treatments as with other organisms.

6. Traits which are of interest to study are often extremely complex and difficult to quantify.

8. Usually, if traits fail to breed true in animals which are *considered to be* genetically homozygous (the pure breeds of dogs) it is likely that the trait is determined by conditioning or complex genetic factors which each have minor influences. Even though the dogs are "pure breeds" it is likely that alleles are segregating and combining in various ways to produce the variations noted. There may also be different interactions with environmental stimuli from such subtle genetic variations.

25

Population Genetics

Vocabulary: Organization and Listing of Terms

Heterozygote frequency

$$\sqrt{q^2}$$

$p = 1 - q$

$2pq$

Demonstrating equilibrium

    expected frequencies

    observed frequencies

    nonequilibrium

Changes in gene frequencies

    mutation (generates variability)

        recessive

        dominant

            achondroplasia

    migration

        Duffy blood group

        selection

            natural selection

                Biston betularia

            fitness

            selection coefficient (*s*)

                directional selection

                stabilizing selection

                disruptive selection

                    Drosophila

                        bristle number

genetic drift

    small populations

    population size

        Drosophila

            forked bristles

    isolated subpopulations

        achromatopsia

        Dunkers

        ABO, MN

inbreeding and heterosis

    assortative

        inbreeding

        self-fertilization

    consanguineous marriages

    coefficient of inbreeding

    inbreeding depression

        hybrid

        heterosis

            dominance theory

            overdominance

Concepts

Population genetics

Gene pool

    gene (allelic) frequencies (F25.1)

Population

Hardy-Weinberg assumptions

    infinitely large

        no drift

    random mating

no selection

no mutation

no migration

Genetic equilibrium

    genetic variability

Inbreeding and heterosis (F25.2)

F25.1. Diagram of the relationships among populations, individuals, alleles, and allelic frequencies (*p*, *q*).

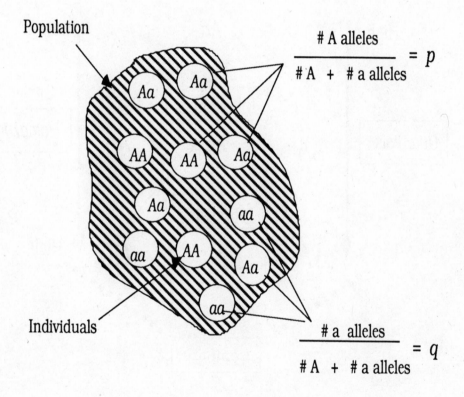

F25.2. Diagram of the relationsips among inbreeding, heterosis, and homozygosity. Note that as inbreeding occurs, heterosis decreases while homozygosity increases.

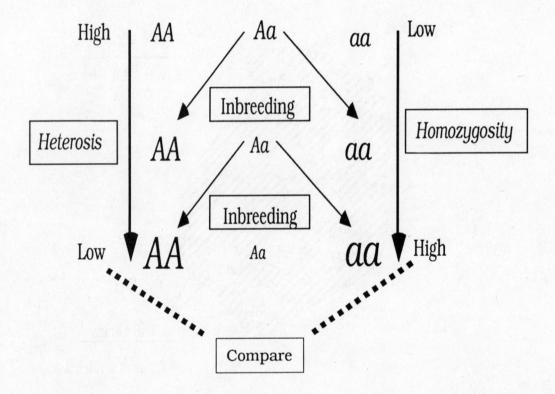

Solutions to Problems and Discussion Questions

1. Because the alleles follow a dominant/recessive mode, one can use the equation $\sqrt{q^2}$ to calculate q from which all other aspects of the answer depend. The frequency of *aa* types is determined by dividing 37 (number of non-tasters) by the total number of individuals (125).

$$q^2 = 37/125 = .296$$

$$q = .544$$

$$p = 1 - q$$

$$p = .456$$

The frequencies of the genotypes are determined by applying the formula $p^2 + 2pq + q^2$ as follows:

Frequency of *AA* $= p^2$

$\qquad\qquad = (.456)^2$

$\qquad\qquad = .208$ or 20.8%

Frequency of *Aa* $= 2pq$

$\qquad\qquad = 2(.456)(.544)$

$\qquad\qquad = .496$ or 49.6%

Frequency of *aa* $= q^2$

$\qquad\qquad = (.544)^2$

$\qquad\qquad = .296$ or 29.6%

When completing such a set of calculations it is a good practice to add the final percentages to be certain that they total 100%.

2. Understanding the Hardy-Weinberg equilibrium allows us to state that if a population is equilibrium, the genotypic frequencies will not shift from one generation to the next unless there are factors such as selection, migration, etc. which alter gene frequencies. Since none of these factors are stated in the problem, we need only to determine whether the initial population is in equilibrium. Calculate p and q, then apply the equation $p^2 + 2pq + q^2$ to determine genotypic frequencies in the next generation.

p $=$ frequency of *A*

$\qquad = 0.2 + .3$

$\qquad = 0.5$

q $= 1 - p = 0.5$

Frequency of *AA* $= p^2$

$\qquad\qquad = (.5)^2$

$\qquad\qquad = .25$ or 25%

Frequency of *Aa* $= 2pq$

$\qquad\qquad = 2(.5)(.5)$

$\qquad\qquad = .5$ or 50%

Frequency of *aa* $= q^2$

$\qquad\qquad = (.5)^2$

$\qquad\qquad = .25$ or 25%

The initial population was not in equilibrium, however after one generation of mating under the Hardy-Weinberg assumptions, the population is in equilibrium and will continue to be so (and not change) until one or more of the Hardy-Weinberg assumptions is not met. Note that *equilibrium* does not necessarily mean p and q equal 0.5.

3. For each of these values, one merely takes the square root to determine q, then one computes p, then one "plugs" the values into the $2pq$ expression.

(a) $q = .08$; $2pq = 2(.92)(.08)$

$= .1472$ or 14.72%

(b) $q = .009$; $2pq = 2(.991)(.009)$

$= .01784$ or 1.78%

(c) $q = .3$; $2pq = 2(.7)(.3)$

$= .42$ or 42%

(d) $q = .1$; $2pq = 2(.9)(.1)$

$= .18$ or 18%

(e) $q = .316$; $2pq = 2(.684)(.316)$

$= .4323$ or 43.23%

(depending how one rounds off the decimals, slightly different answers will occur)

4. In order for the Hardy-Weinberg equations to apply, the population must be in equilibrium.

5. If one has the frequency of individuals with the dominant phenotype, the remainder have the recessive phenotype (q^2). With q^2 one can calculate q and from this value one can arrive at p. Applying the expression $p^2 + 2pq + q^2$ will allow a solution to the question.

6. (a) For the MN analysis, first determine p and q. Since one has the frequencies of all the genotypes, one can add

.6 and .351/2 to provide p (= .7755);

q will be

1-.7755 or .2245.

The equilibrium values will be as follows:

Frequency of *MM* $= p^2 = (.7755)^2$

$= .6014$ or 60.14%

Frequency of *MN* $= 2pq$

$= 2(.7755)(.2245)$

$= .3482$ or 34.82%

Frequency of *NN* $= q^2 = (.2245)^2$

$= .0504$ or 5.04%

Comparing these equilibrium values with the observed values strongly suggest that the observed values are drawn from a population in equilibrium.

(b) For the *AS* analysis, first determine p and q. Since one has the frequencies of all the genotypes, one can add

.756 and .242/2 to provide p (= .877);

q will be

1-.877 or .123.

The equilibrium values will be as follows:

Frequency of AA = p^2 = $(.877)^2$

= .7691 or 76.91%

Frequency of AS = $2pq$= $2(.877)(.123)$

= .2157 or 21.57%

Frequency of SS = q^2 = $(.123)^2$

= .0151 or 1.51%

Comparing these equilibrium values with the observed values suggest that the observed values may be drawn from a population which is not in equilibrium. Notice that there are more heterozygotes than predicted, and fewer SS types. To test for a Hardy-Weinberg equilibrium, apply the chi-square test as follows.

$$\chi^2 = \sum \frac{(o-e)^2}{e}$$

$(75.6-76.9)^2/76.9$ +

$(21.6 - 24.2)^2/24.2$ +

$(1.51 - 0.2)^2/0.2$ =   8.88

Checking the χ^2 table with 2 degrees of freedom (number of classes - 1) gives a value of 5.99 at the 0.05 probability level. Since the χ^2 value calculated here is greater, the null hypothesis (the observed values fluctuate from the equilibrium values by chance and chance alone) should be rejected. Thus the frequencies of AA, AS, SS sampled a population which is not in equilibrium.

7. Given that q^2= .04, then q = .2, $2pq$ = .32, and p^2= .64 which is the frequency of heterozygotes in the population. Of those not expressing the trait, only a mating between heterozygotes can produce an offspring which expresses the trait, and then only at a frequency of 1/4. The different types of matings possible (those without the trait) in the population, with their frequencies are given below:

AA X AA = .64 X .64 = .4096

AA X Aa = .64 X .32 = .2048

Aa X AA = .64 X .32 = .2048

Aa X Aa = .32 X .32 = .1024

Aa X Aa = .32 X .32 = .1024

Notice that of the matings of the individuals who do not express the trait, only the last two (about 20%) are capable of producing offspring with the trait. Therefore one would arrive at a final likelihood of 1/4 X 20% or 5% of the offspring with the trait.

8. $q_1 = (q_o - sq_o^2)/(1 - sq_o^2)$,

with p = .7 and q = .3

and s being the selection coefficient,

(a) q_1 = [.3 - (1.0 X .09)]/[1 - (1.0 X .09)]

q_1 = .23
p_1 = .77

(b) q_1 = [.3 - (.5 X .09)]/[1 - (.5 X .09)]

q_1 = .267
p_1 = .733

(c) $q_1 = [.3 - (.1 \text{ X } .09)]/[1 - (.1 \text{ X } .09)]$

$q_1 = .293$
$p_1 = .707$

(d) $q_1 = [.3 - (.01 \text{ X } .09)]/[1 - (.01 \text{ X } .09)]$

$q_1 = .299$
$p_1 = .701$

9. The general equation for responding to this question is

$$q_n = q_o/(1 + nq_o)$$

where n = the number of generations, q_o = the initial gene frequency, and q_n = the new gene frequency.

(a)

$q_n = q_o/(1 + nq_o)$

$q_n = 0.5 / [1 + (1 \text{ X } 0.5)]$

$q_n = .33$

(b)

$q_n = q_o/(1 + nq_o)$

$q_n = 0.5 / [1 + (5 \text{ X } 0.5)]$

$q_n = .143$

(c)

$q_n = q_o/(1 + nq_o)$

$q_n = 0.5 / [1 + (10 \text{ X } 0.5)]$

$q_n = .083$

(d)

$q_n = q_o/(1 + nq_o)$

$q_n = 0.5 / [1 + (25 \text{ X } 0.5)]$

$q_n = .037$

(e)

$q_n = q_o/(1 + nq_o)$

$q_n = 0.5 / [1 + (100 \text{ X } 0.5)]$

$q_n = .0098$

(f)

$q_n = q_o/(1 + nq_o)$

$q_n = 0.5 / [1 + (1000 \text{ X } 0.5)]$

$q_n = .00099$

10. For this question, apply the equations

$$\Delta p = m(p_m - p)$$

and $p_1 = p + \Delta p.$

Substituting, gives: $p_1 = p + m(p_m - p).$

(a) $p_1 = 0.6 + 0.2(0.1 - 0.6) = 0.5$

(b) $p_1 = 0.2 + 0.3(0.7 - 0.2) = 0.35$

(c) $p_1 = 0.1 + 0.1(0.2 - 0.1) = 0.11$

11. Carefully examine K/C-Fig.25.12. What one must do is predict the probability of one of the grandparents being heterozygous in this problem. Given the frequency of the disorder in the population as 1 in 10,000 individuals (0.0001), then q^2 = 0.0001, and q = 0.01. The frequency of heterozygosity is *2pq* or approximately .02 as also stated in the problem. The probability for one of the grandparents to be heterozygous would therefore be 0.02 + 0.02 or 0.04 or 1/25.

If one of the grandparents is a carrier, then the probability of the offspring from a first-cousin mating being homozygous for the recessive gene is 1/16. Multiplying the two probabilities together gives 1/16 X 1/25 = 1/400.

Following the same analysis for the second-cousin mating gives 1/64 X 1/25 = 1/1600. The population at large has a frequency of homozygotes of 1/10,000, therefore one can easily see how inbreeding increases the likelihood of homozygosity.

12. *Inbreeding depression* refers to the reduction in fitness observed in populations which are inbred. With inbreeding comes an increase in the number of homozygous individuals (see F24.2 in this book) and a decrease in genetic variability. Genetic variability is necessary for a genetic response to environmental change. As deleterious genes become homozygous, more individuals are less fit in the population.

13. While inbreeding increases the frequency of homozygous individuals in a population, it does not change the *gene* frequencies. There will be fewer heterozygotes in the population to compensate for the additional homozygotes. See K/C-Fig.25.11 and F24.2 in this book.

14. Because heterozygosity tends to mask expression of recessive genes which may be desirable in a domesticated animal or plant, inbreeding schemes will often be used to render strains homozygous so that such recessive genes can be expressed.

In addition, assume that a particularly desirable trait occurs in a domesticated plant or animal. The best way to increase the frequency of individuals with that trait is by self-fertilization (not often possible) or by matings to blood relatives (inbreeding). In theory, one increases the likelihood of a gene "meeting itself" by various inbreeding schemes. There are disadvantages to increasing the degree of homozygosity by inbreeding. *Inbreeding depression* is a reduction in fitness often associated with an increase in homozygosity.

15. While inbreeding increases the frequency of homozygous individuals in a population, it does not change the *gene* frequencies. There will be fewer heterozygotes in the population to compensate for the additional homozygotes. See K/C-Fig.25.11 and F24.2 in this book.

16.

(a) The quickest way to generate a homozygous line of an organism is to *self-fertilize* that organism. Because this is not always possible, brother-sister matings are often used.

(b) Notice in K/C-Fig.25.11 that with self-fertilization, the percentage of homozygous individuals increases dramatically.

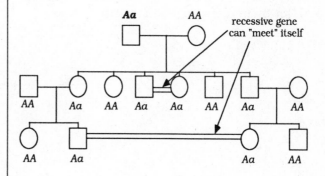

17. Given that the recessive gene a is present in the homozygous state (q^2) at a frequency of 0.0001, the value of q is 0.01 and $p = 0.99$.

(a) q is 0.01

(b) $p = 1 - q$ or .99

(c) $2pq = 2(.01)(.99)$

$= 0.0198$ (or about 1/50)

(d) $2pq \times 2pq$

$= 0.0198 \times 0.0198$

$= 0.000392$ or about 1/255

18. The frequency of a gene is determined by a number of factors including the fitness it confers, mutation rate, and input from migration. There is no tendency for a gene to reach any artificial frequency such as 0.5. The distribution of a gene among individuals is determined by mating (population size, inbreeding, etc.) and environmental factors (selection, etc.). A population is in equilibrium when the distribution of genotypes occurs at or around the $p^2 + 2pq + q^2$ expression.

Equilibrium does not mean 25% *AA*, 50% *Aa*, and 25% *aa*. This confusion often stems from the 1:2:1(or 3:1) ratio seen in Mendelian crosses.

19. Because three of the affected infants had affected parents, only two "new" genes, from mutation, enter into the problem. The gene is dominant, therefore each new case of achondroplasia arose from a single new mutation. There are 50,000 births, therefore 100,000 genes involved. The frequency of mutation is therefore given as follows:

2/100,000

or 2×10^{-5}

26

Evolutionary Genetics

Vocabulary: Organization and Listing of Terms

Structures and Substances

Allozyme

Insulin

Cytochrome c

Carbonic anhydrase

Myoglobin

Hemoglobin

Processes/Methods

Evolutionary divergence (F26.1)

Genetic divergence

Genetic diversity

   heterozygosity

     protein polymorphism

       allozymes (F26.2)

         molecular phylogenetic trees

amino acid sequence homology

   inslulin

   cytochrome c

   carbonic anhydrase

   myoglobin

   hemoglobin

chromosomal polymorphism

   inversions

   translocations

DNA sequence polymorphism

    (phylogeny)

  nucleic acid hybridization

   homologous reaction

   heterologous reaction

   ΔT_m

   Strongylocentrotus sp.

   Drosophila sp.

RFLP phylogenies

Homo sp.

mitochondrial

nuclear

gene duplication

pseudogenes

orphons

Inbreeding depression

deleterious alleles

Drosophila sp.

Ecological diversity

niche (F26.1)

Preadaptation

Speciation

phyletic evolution

geographic or allopatric speciation

geographic barriers

isolating mechanisms

reproductive

postzygotic

prezygotic

Drosophila sp.

sibling species

gradualism

quantum

stochastic (catastrophic)

punctuated equilibria

evolutionary spurts

founder-flush

flush

population crash

Hawaiian *Drosophila sp.*

statispatric

homokaryotypes

allopolypolody

Nicotiana sp.

Gel electrophoresis

Concepts

Evolutionary divergence

minimal mutational distance

divergence dendrograms

(phylogenetic tree)

Species formation (speciation)

species

interbreeding

potentially interbreeding

Race (ecotype)(subspecies) (F26.1)

Niche

Adaptive norm

   preadaptation

Genome size

   classes of genome size

Mutation and speciation

   classical hypothesis

   balance hypothesis

neutralist hypothesis

   functional constraints (F26.3)

   cost of selection

   selectionists

Recent trends

   genome in flux

   transposable elements

   mutations in development

F26.1. Diagram of possible relationships among several aspects of race formation. Different subgroups of a population which differ genetically are illustrated in the figure as separate races each occupying a different niche.

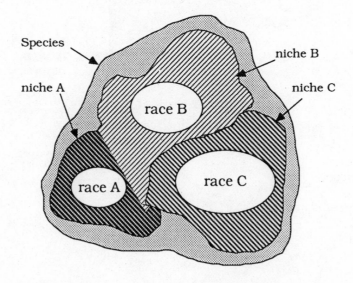

F26.2. The diagram below is meant to illustrate the meaning of the term *allozyme*. Notice that alleles genes A' and A'' produce protein products which differ electrophoretically but are involved in the same function.

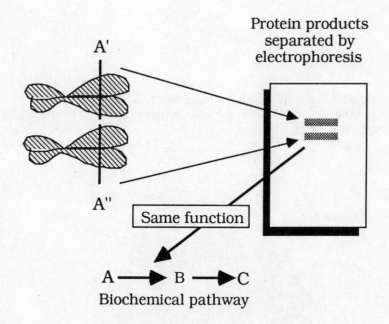

F26.3. The illustration below is meant to show that certain parts of the protein would be more able to withstand changes in amino acid sequence than those with functional constraints. Those areas most distant from the "functional portions" would seem to be more tolerant to change.

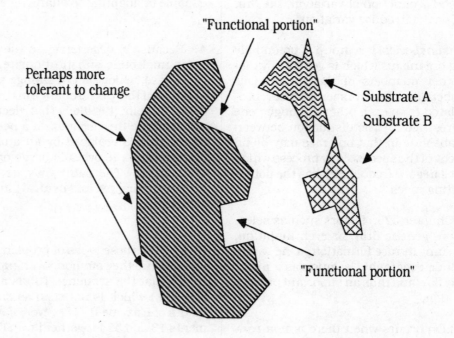

Solutions to Problems and Discussion Questions

1. Assume that a chromosome in the "standard arrangement" (see K/C-Fig.26.6) undergoes an inversion (pericentric or paracentric). The following are possible consequences of such an inversion:

(a) change in gene order with possible introduction of position effects,

(b) breakage within a structural gene or other functional element,

(c) reduction in the recovery of crossover gametes in heterokaryotypes (those which carry an inversion as well as a standard homologue). While "c" may reduce the production of variation, the first two (a and b) may introduce variation.

2. A *race* or *subspecies* is a group of potentially interbreeding organisms which is genetically distinct from other members of the species (see F26.1). Members of different races are not reproductively isolated from each other although gene flow may be restricted. The distinction between races is not absolute in that one race may blend with other races of the species. Any process which favors changes in gene frequencies has the potential of generating races.

As stated in Chapter #25, factors such as selection, migration, genetic drift, or even mutation, may be important in race formation. One would certainly include geographic isolation as a major barrier to gene flow and thus an important process in race formation.

Natural selection occurs when there is non-random elimination of individuals from a population. Since such selection is a strong force in changing gene frequencies, it should also be considered as a significant factor in race formation.

3. The *classical hypothesis* supports the notion that the very nature of natural selection favors genotypes which are the most well adapted to a given environment. Therefore, each organism in a given environment should possess the most favorable genotype and possess little genotypic variation.

The *balance hypothesis* favors a model in which significant genetic variation (heterozygosity) is maintained in a population and all organisms are not ideally suited to each environment. Dobzhansky described an *adaptive norm* (an array of genotypes in a population) in terms of a balance between selection against extreme phenotypic variations and the maintenance of genetic heterozygosity. A successful population in both space and time is one which has sufficient genetic variability to produce a variety of phenotypes capable of adapting to changing environments.

4. Because of degeneracy in the code, there are some nucleotide substitutions, especially in the third base, which do not change amino acids. In addition, if there is no change in the overall charge of the protein, it is likely that electrophoresis will not separate the variants. If a positively charged amino acid is replaced by an amino acid of like charge, then the overall charge on the protein is unchanged. The same may be said for other, negatively charged and neutral amino acid substitutions.

5. To solve these types of problems, one must try to match up the common, overlapping sequences. Notice that the sequence 135 is a fragment from Enzyme I which is not seen as such in the fragments of Enzyme II. However, if one places fragments 13 and 524 together (from Enzyme II), then the sequence 13524 is generated. Also notice that other pieces can be placed together, such as 135246356 (from Enzyme II) will match with 13524635 from the fragments of Enzyme I. One can finish off the sequence by adding the last "624" from Enzyme I for the complete sequence: 13524635624.

6. All of the amino acid substitutions

(Ala -> Gly, Val -> Leu, Asp -> Asn, Met -> Leu)

require only one nucleotide change. The last change from

Pro (CC-) -> Lys (AAA,G)

requires two changes (the minimal mutational distance).

7. Approach this problem by writing the possible codons for all the amino acids (except Arg and Asp which show no change) in the human cytochrome c chain. Then determine the minumum number of nucleotide substitutions required for each changed amino acid in the various organisms. Once listed, then count up the numbers for each organism: horse, 3; pig, 2; dog, 3; chicken, 3; bullfrog, 2; fungus, 6.

8. The classification of organisms into different species is based on evidence (morphological, genetic, ecological, etc.) that they are reproductively isolated. That is, there must be evidence that gene flow does not occur among the groups being called different species. Classifications above the species level (genus, family, etc.) are not based on such empirical data. Indeed, classification above the species level is somewhat arbitrary and based on traditions which extend far beyond DNA sequence information.

In addition, recall that DNA sequence divergence is not always directly proportional to morphological, behavioral, or ecological divergence. Therefore, while the genus classifications provided in this problem seem to be invalid, other factors, well beyond simple DNA sequence comparison, must be considered in classification practices. As more information is gained on the meaning of DNA sequence differences (ΔT_m) in comparison to morphological factors, many phylogenetic relationships will be reconsidered and it is possible that adjustments will be needed in some classification schemes.

9. In looking at the figure, notice that the $\Delta T_{50}H$ value of 4.0 on the right could be used as a decision point such that any group which diverged above that line would be considered in the same genus while any group below would be in a different genus. Under this rule, one would have the chimpanzee, pygmy chimpanzee, human, gorilla and orangutan in the same genus. If one assumed that 3.7 is close enough to be considered above 4.0, given considerable experimental error, one could provide a scheme where the orangutan is not included with the chimpanzee, pygmy chimpanzee, human, and gorilla.

10. There are many sections of DNA in a eukaryotic genome which are not reflected in a protein product. Indeed, there are many sections of DNA which are not even transcribed and/or have no apparent physiological role. Such regions are more likely to tolerate nucleotide changes compared to those regions with a necessary physiological impact. Introns for example show sequence variation which is not reflected in a protein product. Exons on the other hand code for products which are usually involved in production of a phenotype and as such are subject to selection.

11. The text lists several cornerstones of the *neutral mutation theory*:

(a) the relatively uniform rate of amino acid substitution in different organisms (under different types of selection),

(b) there is no particular pattern to the substitutions indicating that selection is not eliminating some variations,

(c) the rate of mutation is relatively high and has remained relatively constant for millions of years even though environments have fluctuated greatly over that period of time,

(d) certain regions of molecules and certain functions of those molecules should logically be less likely to have amino acid substitutions influence the phenotype,

(e) the rate of amino acid substitution in some proteins is much too high to have been produced by selection. The *selectionists* suggest that even though amino acid substitutions *appear* to be neutral, it is more likely that their influence has just not been determined. In addition, they point out that many polymorphisms are clearly maintained in the population *by* selection.

12. Like many other debates which surround the nature of evolution, it is important to see that debate is a natural component of scientific understanding. It is likely that some genes (like histones) will not tolerate nucleotide substitutions to a significant degree and the neutral mutation theory will not hold. However, there are other genes which produce quite variable products and provide support for the neutral mutation theory. Usually controversy is resolved as one dives deeper into the problem and seeks to define the variables and complexities of the process. It is controversy which stimulates a desire to seek answers.

Part Four: Sample Test Questions (with detailed explanations of answers)

Question 1. Two terms, *determination* and *differentiation*, are consistently used in discussions of development.

(a) Provide a brief definition of each term.

(b) Which, determination or differentiation, comes first during development of *Drosophila*, for example?

(c) In what way do the two phenomena, *transdetermination* and *regeneration*, provide insights into the processes of determination and differentiation?

Concepts:

> **terms**
>
> > **determination**
> >
> > **differentiation**
>
> **relationships**
>
> > **transdetermination**
> >
> > **regeneration**
>
> ***Drosophila* development**

Answer 1. **(a)** *Determination* is a significant, complex, yet poorly understood process whereby the specific pattern of genetic activity is initially established in a cell. This pattern will direct the developmental fate (differentiation) of that cell. *Differentiation* is the process of cellular expression of the determined state. It is the complex series of genetic, morphological, and physiological changes which characterize the variety of adult cells.

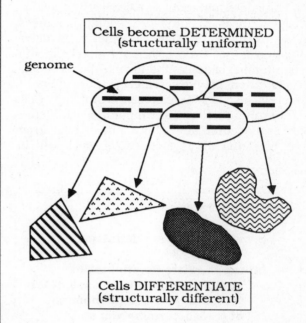

(b) *Determination* occurs before *differentiation*. In *Drosophila*, determinative events are thought to occur about the time of blastoderm formation (see Chapter 21 of the K/C text) when nuclei encounter the peripheral regions of the egg. *Differentiation* of most of the adult cells occurs during metamorphosis, some five to six days after embryogenesis (determination).

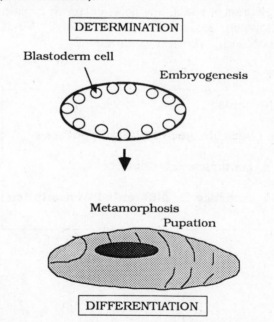

(c) Transdetermination, revealed by the culturing imaginal disks in adult *Drosophila* abdomens, indicates that under some circumstances, determination is not fixed. Thus the developmental programming of determination may be reversible. Regeneration provides evidence that differentiated cells also lack developmental stability. Cells in the stump portion of an amputation dedifferentiate and accumulate to form a blastema. From that blastema all regenerating tissues will be derived.

Common errors:

> **providing accurate definitions**
>
> **failure to relate determination and differentiation to each other temporally and to phenomena of transdetermination and regeneration**

Question 2. Development may be defined as the attainment of a differentiated state. Given that all cells of a eukaryote probably contain the same complete set of genes, how do we currently explain development in terms of gene activity? What evidence supports your explanation?

Concepts:

> **variable gene activity hypothesis**
>
> **genomic equivalence**
>
> **evidence for differential transcription**

Answer 2. The *variable gene activity hypothesis* of differentiation acknowledges the genomic equivalence of cells within an organism and assumes that of all the genes in a given cell type only certain ones produce products while the others are shut down and are not transcribed. As shown below certain genes will be active in all cells, those *housekeeping* genes coding for vital cellular functions, while others will be differentially regulated in various cell types. Differential gene transcription occurs in both spatial (different cells of an organism) and temporal (different times during development) dimensions.

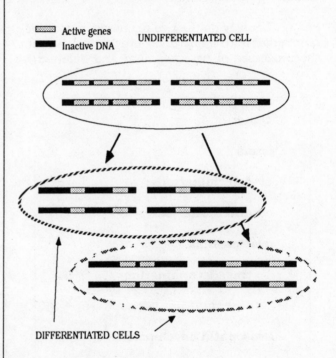

Support for this model is provided by several observations and experiments.

1. *Chromosome puffs:* Specific puff patterns, representing differential gene activity, are observed in dipteran polytene chromosomes at different times during development.

2. *Isozymes:* Differential gene activity is demonstrated by the observation that different forms of the same enzyme (isozymes) are present in cells of different tissues. This evidence assumes that such isozyme patterns are not caused by post-transcriptional forms of genetic regulation.

3. *Growth hormone:* *In situ* hybridization and immunochemical studies in mouse embryos demonstrates spatial and temporal aspects of the regulation of growth hormone transcripts in the anterior pituitary gland.

4. *Bacterial sporulation:* Analysis of developmental mutations in spore-forming bacteria indicates that differential gene activity is regulated in part by different forms of RNA polymerase.

Common errors:

With a general question such as this students sometimes have difficulty focusing on the area in their notes or in the text which relates to the question. Students may understand what is meant by the *variable gene activity hypothesis* but not immediately see that it relates to the question. Students also have difficulty in relating a variety of experimental findings to a general theme.

Question 3. Direction of shell coiling in the land snail, *Limnaea peregra*, is determined by alleles at a single locus: *dextral* (right) = *DD* or *Dd*; *dd* = *sinistral* (left). However, a maternal effect is present such that the genotype of the mother determines the direction of coiling (phenotype) of the immediate offspring. Given the following crosses, write the genotypes and phenotypes in the spaces provided. Be certain to indicate which genotypes go with which phenotypes.

	Source of sperm	*Source of egg*
Cross #1	**DD**	**dd**

Offspring genotype(s): *Offspring phenotype(s):*

Cross #2	**dd**	**Dd** (sinistral)

Offspring genotype(s): *Offspring phenotype(s):*

Cross #3	**Dd**	**Dd**

Offspring genotype(s): *Offspring phenotype(s):*

Concepts

maternal effects

Answer 3: The definition provided in the initial problem can be applied directly to the solution of this problem. The *genotype* of the mother determines the direction of coiling (phenotype) of the immediate offspring. In this case, one merely assigns the genotypes on the basis or normal Mendelian principles, then the phenotypes based on the genotype of the mother.

Notice in Cross 2, the *Dd* has a sinistral phenotype. While this may confuse some students, remember that the phenotype is determined by the *maternal* genotype. When early developmental events are involved, often the maternal genotype will have a significant influence over those events because the mother makes the egg.

Cross #1:

Offspring genotype(s):	Offspring phenotype(s):
Dd	all sinistral

Cross #2:

Offspring genotype(s):	Offspring phenotype(s):
Dd *dd*	all dextral

Cross #3:

Offspring genotype(s):	Offspring phenotype(s):
DD *Dd* *dd*	all dextral

Common errors:

When students are reminded of the nature of a maternal effect, that is, given a definition, there are very few errors. However, when they are asked to work the problem without the definition given, many have difficulty. Cross #2 causes most of the problems.

Question 4. Extrachromosomal inheritance is a general term which includes maternal effects, organelle heredity, and infectious heredity. Provide a short description and example of each. How would one distinguish between a maternal effect and a case of organelle heredity?

Concepts:

   extrachromosomal inheritance

   categories

   examples

   experimental approaches

Answer 4: A *maternal effect* is a case of extrachromosomal inheritance in which the genotype of the mother influences the phenotype of her immediate offspring in a non-Mendelian manner. Maternal effects are thought to be caused by the mother providing molecules of various types (substrates, products, enzymes, cofactors, mRNAs, etc.) for early development. An example of a maternal effect is the classic one of the direction of shell coiling in the snail *Limnaea*.

Organelle heredity on the other hand is dependent on cytoplasmic components, organelles such as mitochondria and/or chloroplasts, which have genetic properties of their own. Because the egg contains more cytoplasm, therefore organelles, than the sperm or pollen, a non-Mendelian pattern of inheritance is observed. A typical example involves the *poky* mutation in *Neurospora*.

In *infectious heredity* some infective agent (virus, protozoan) inhabits the egg and is passed from the "egg-parent" to the offspring in a non-Mendelian manner. An example is the infective virus, *sigma*, in *Drosophila*.

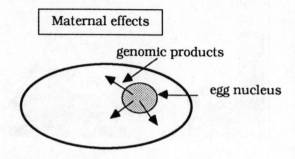

Maternal effects

genomic products

egg nucleus

Organelle or
Infectious heredity

mitochondria, chloroplasts,
or infectious particles

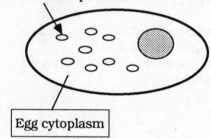

Egg cytoplasm

To distinguish between a maternal effect and a case of organelle heredity, the following approach may be used. Remember that a maternal effect persists for only one generation whereas in organelle heredity the agent persists for many generations as long as the maintenance of the organelle is not dependent on the nuclear genotype.

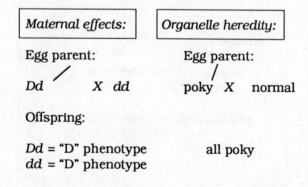

Maternal effects:	*Organelle heredity:*
Egg parent:	Egg parent:
Dd X *dd*	poky X normal
Offspring:	
Dd = "D" phenotype	all poky
dd = "D" phenotype	

But now take the *dd* type which is phenotypically "D" and, as shown below, the phenotype only lasts for one generation.

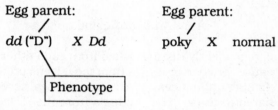

Egg parent:	Egg parent:
dd ("D") X *Dd*	poky X normal
Phenotype	

Offspring:

Dd = "d" phenotype	all poky
dd = "d" phenotype	

Common errors:

Students are generally able to provide definitions and examples, especially when the information is clearly stated in the text. The major source of difficulty for students is the set up an experiment to demonstrate something. More often than not, they will leave this "experimental" aspect blank on an examination, or repeat some experiment from their lecture notes or the text which will not apply to the question.

Question 5. Describe similarities and differences between *discontinuous* and *continuous* traits at the *molecular* and *transmission* levels.

Concepts:

interaction of gene products

relationship between genotype and phenotype

multi-factor inheritance

Answer 5. At the *molecular level* one could consider that in discontinuous inheritance the gene products are acting fairly independently of each other, thereby providing a 9:3:3:1 ratio in a dihybrid cross for example. Genotypic classes

A-B-, A-bb, aaB-, and *aabb*

can be clearly distinguished from each other because the gene products from the *A* locus produce distinct influences on the phenotypes as compared to those gene products from the *B* locus. Exceptions exist where epistasis and other forms of gene interaction occur. In discontinuous inheritance one would consider each locus as providing a *qualitatively* different impact on the phenotype. For instance, even though the *brown* and *scarlet* loci interact in the production of eye pigments in *Drosophila*, each locus is providing qualitatively different input.

In continuous inheritance, we would consider each involved locus as having a quantitative input on the production of a single characteristic of the phenotype. In addition, although it may not always be the case, we would consider each gene product as being qualitatively similar. Under this model, the *quantity* of a particular set of gene products, influenced by a number of gene loci, determines the phenotypic characteristic.

At the *transmission level* one sees "step-wise" distributions in discontinuous inheritance but "smoother" or more bell-shaped distributions in continuous inheritance as shown in the figure below. For instance, in a dihybrid situation (*AaBb X AaBb*) where independent assortment holds, one would obtain a 9:3:3:1 ratio (assuming no epistasis, etc.) under a discontinuous mode but a 1:4:6:4:1 where genes (or gene products) are acting additively (continuous inheritance). Both patterns are formed from normal Mendelian principles of segregation, independent assortment, and random union of gametes. It is the manner in which the genes (or gene products) interact which distinguishes discontinuous from continuous inheritance.

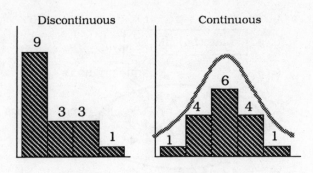

Common errors:

    difficulty with "molecular level' of the question

    confusion over differences and similarities relating discontinuous and continuous patterns

---

Question 6: Phenotypic expression is a complex process involving many levels (including regulatory, metabolic, and developmental) of cellular activity. Two terms, *penetrance* and *expressivity*, are often used to describe phenotypic expression. Provide a definition of each of these terms, then describe factors which vary *penetrance* and *expressivity*.

Concepts:

    "global" view of phenotypic expression

    relationship of penetrance and expressivity to other factors

Answer 6: *Penetrance* may be defined as the percentage of individuals that express a mutant genotype. Expression may be complete or partial. *Expressivity* refers to the degree of expression of a given genotype. A variety of factors are known to influence penetrance and expressivity. Such factors are expected to act at the levels of gene regulation, transcription, translation, and "higher" developmental processes.

Genetic background. While it is often difficult to assess the influence of genetic background on gene expression, many genes are influenced by "modifiers" which enhance or suppress expression. Since each individual gene functions within an environment produced by all other active genes of the genome, it is reasonable to expect that a host of "background" factors will influence the expression of a gene. Such background factors may involve positional as well as molecular/developmental factors.

Temperature. Since biochemical activity depends on levels of kinetic energy and gene activity and expression are somewhat "biochemical," it is natural to suspect that temperature will influence phenotypic expression.

Nutrition. The raw materials used in growth and maintenance of the living state are usually provided by the diet. As nutrients fluctuate, changes will be expected in the metabolic state of the organism. Such changes may influence the series of biochemical reactions involved in the expression of a gene.

Common errors:

definitions of the penetrance and expressivity

relating the two terms to "environmental factors"

listing and briefly describing the terms

Question 7: Assume that in a particular population, approximately 8% of the males show red-green color blindness. Knowing that this form of color blindness is X-linked, what percentage of the females would be expected to be color blind?

What would be the expected frequency of heterozygous females?

Assuming that the Hardy-Weinberg equilibrium assumptions pertain, what percentage of men will be color blind in the next generation?

Concepts:

Hardy-Weinberg applications to X-linked gene

maintenance of gene frequencies over time

Answer 7: Since 8% of the males express the trait and males have only one X chromosome, the frequency (q) of the recessive gene would be .08 and p would be .92. Since females have two X chromosomes, the expected frequency of females that are homozygous for the color blindness gene would be q^2 or .0064 (.64%).

The frequency of females that are heterozygous would be $2pq$ or $2(.08)(.92) = .1472$ or 14.72%.

Because the population is in equilibrium, the frequency of men with color blindness will not change from generation to generation. Eight percent of the men will be color blind in the next generation.

Students should be aware of many deviations of these types of questions. The basic scheme is the Hardy-Weinberg equilibrium and the equations which apply.

Common errors:

> **application of the Hardy-Weinberg equations to a sex-linked gene**
>
> **failure to apply the Hardy-Weinberg equations in determining the frequency of heterozygous females**
>
> **students often make the question harder than it is by forgetting that under equilibrium conditions, gene frequencies do not change**

Question 8: List and briefly describe factors which change gene frequencies in populations. Is inbreeding a factor in changing gene frequencies? Explain.

Concepts:

> **factors which change gene frequencies**
>
> **influence of inbreeding on gene frequencies**

Answer 8:

Mutation, while being an original source of genetic variability is not usually considered to be a significant factor in changing gene frequencies.

Migration occurs when individuals move from one population to another. The influence of migration on changing gene frequencies is proportional to the differences in gene frequency between the donor and recipient populations.

Organisms often migrate as a result of some stress. Those organisms suffering from the most stress are often those which leave. Therefore, they do not represent a random sample of the individuals in that home range.

Selection can be a significant force in changing gene frequencies. It results when some genotypic classes are less likely to produce offspring than others. Selection may be directional, stabilizing, or disruptive.

Genetic drift can be a significant force in changing gene frequencies in populations which are numerically small or have a small number of effective breeders. In such populations, random and relatively large fluctuations in gene frequency occur by "sampling error."

Inbreeding is not a significant factor in changing gene frequencies in populations however it will change zygotic or genotypic frequencies. The number of homozygotes will increase at the expense of the heterozygotes.

Common errors:

> **failure to provide a complete list**
>
> **failure to briefly and adequately describe each term**
>
> **failure to see that inbreeding does not, in itself, change gene frequencies**

Question 9: A *species* is often defined as a population of interbreeding or potentially interbreeding organisms which is reproductively isolated from other such populations. Given such an isolated population, will speciation (formation of a new species) occur if the Hardy-Weinberg assumptions are met?

(c) Twin studies have greatly facilitated our understanding of such human behavioral traits as schizophrenia and manic-depression. Why?

(d) What factors are believed to influence the expressivity of human behaviors?

Concept:

relationship between speciation and Hardy-Weinberg assumptions

---

Answer 9: Reproductive isolation can occur because of the introduction of geographic barriers or other dramatic changes in the environment which subdivide a population. Such factors facilitate speciation because gene flow is eliminated or at least restricted. With gene flow restricted, isolated populations can experience changes in gene frequencies when the Hardy-Weinberg assumptions are *not* met. Under Hardy-Weinberg equilibrium conditions where there is *random mating, no genetic drift, no selection, no mutation,* and *no migration,* gene frequencies will remain the same and speciation will not occur.

Common errors:

confusion as to what the question is asking

difficulty in relating information from one chapter to information contained in a different chapter

---

Question 10. The study of behavioral genetics is complicated by the fact that many behavioral traits of interest are determined by more than one gene pair and often characterized by variation in penetrance and expressivity. The study of human behavioral genetics suffers additional complications.

(a) List three of these additional complications.

(b) The neurological disorder of Huntington's disease is caused by an autosomal dominant gene yet in pedigrees it may show incomplete penetrance. Why?

Concepts:

experimental approaches to human genetics

twin studies

penetrance, expressivity

Answer 10:

(a) Several additional problems in the study of human behavioral genetics would include the following.

1. With a relatively small number of offspring produced per mating, standard genetic methods of analysis are difficult.

2. Records on family illnesses, especially behavioral illnesses, are difficult to obtain.

3. The long generation time makes longitudinal (transmission genetics) studies difficult.

4. The scientist can not direct matings which will provide the most informative results.

5. The scientists can't always subject humans to the same types of experimental treatments as with other organisms.

(b) With the relatively late age of onset of Huntington's disease, an individual may have children with Huntington's disease, thus indicating the presence of the gene, but die of other causes (accident, military, etc.) before the disease manifests itself. A pedigree having incomplete penetrance will result.

(c) With twin studies, one can readily compare the differential influences of genetic make-up and the environment. Monozygotic twins have the same genetic makeup but can be reared under different environmental conditions if the twins are separated for one reason or another. Dizygotic twins have a different genetic makeup but because they are the same age (almost) they are often reared under fairly uniform conditions within a family. Given these "experimental" advantages, the genetic contribution to certain diseases can often be estimated.

(d) *Genetic background.* While it is often difficult to assess the influence of genetic background on gene expression, many genes are influenced by "modifiers" which enhance or suppress expression. Since each individual gene functions within an environment produced by all other active genes of the genome, it is reasonable to expect that a host of "background" factors will influence the expression of a gene. Such background factors may involve positional as well as molecular/developmental factors.

Environment and *general health.* The physical and behavioral environment in which an individual is raised will influence the degree to which a particular genotype is expressed. Seeing it another way, what may be acceptable or normal behavior in one environment may be abnormal in another. When a living system is stressed by disease or other factors, the chemical/physiological state of the organisms is altered. It is expected that such alterations would be reflected in various behaviors.

Nutrition. The raw materials used in growth and maintenance of the living state are usually provided by the diet. As nutrients fluctuate, changes will be expected in the metabolic state of the organism. Such changes may influence the series of biochemical reactions involved in the expression of a gene. Certain behaviors and known to be influenced by the metabolic state of the individual, thus such behaviors may be influenced by nutrition.

Common errors:

difficulty in relating concepts from different chapters in the text

confusion over penetrance and expressivity